BIOINFORMATICS BASICS

Applications in Biological Science and Medicine

Hooman H. Rashidi, M.S.
Lukas K. Buehler, Ph.D.

CRC Press
Boca Raton London New York Washington, D.C.

Library of Congress Cataloging-in-Publication Data

Catalog record is available from the Library of Congress.

This book contains information obtained from authentic and highly regarded sources. Reprinted material is quoted with permission, and sources are indicated. A wide variety of references are listed. Reasonable efforts have been made to publish reliable data and information, but the author and the publisher cannot assume responsibility for the validity of all materials or for the consequences of their use.

Visit our Website at www. crcpress.com.

No claim to original U.S. Government works
International Standard Book Number 0-8493-2375-4
Printed in the United States of America   4 5 6 7 8 9 0
Printed on acid-free paper

Contents

Foreword

Bioinformatics is an emerging and rapidly growing area of science. Not surprisingly, there is tremendous lag time between the state of knowledge and the publication of manuals for those who need to fathom the basic knowledge and apply it in problem solving. Mr. Rashidi and Dr. Buehler have found a way to address these needs. Their book provides excellent, timely, and easy-to-understand information and advice for those without a great deal of knowledge about bioinformatics. However, the book is far from being simplistic and even those experienced in bioinformatics will find it a handy reference. I personally found the material to be educational and very useful in my own research. I am enthusiastic about the book and confident that it will appeal to a wide range of people, from students to principal investigators.

Dr. Elzbieta Izbicka
Cancer Therapy and Research Center
San Antonio, Texas
and
Adjunct Faculty of Medicine
University of Texas Health Science Center
San Antonio

Preface

After years of research in structure–function relationships, the last decade proved to be tremendously satisfying due to its technical advances in genome sequencing (genomics) and protein identification (proteomics). The existence of public databases with billions of data entries requires a robust analytical approach to cataloging and representing this data with respect to its biological significance. The unifying theme is biological evolution as described by Darwin's theory of evolution and its modern "synthesis" of molecular evolutionary processes. The tool needed to handle this vast amount of data is bioinformatics, and the exponential increase in both computer processing power and disk storage has been instrumental in this age of genes and biotechnology.

Since the idea for this book originated about a year ago, many of the techniques described here have evolved and are constantly changing on the Internet. For example, we refer to the Brookhaven National Laboratory version of the Protein Data Bank. It has been a reliable source of structural information for both proteins and nucleic acids for many years, but by the time this book is printed, the Protein Data Bank will have a new home, the Research Collaboratory for Structural Bioinformatics (RCSB).

Because of today's rapid changes in bioinformatics and Internet-accessible public databases, books like ours often refer to old links. However, the major national and international databases discussed here will be viable for many years. Furthermore, the approach and methodology of bioinformatics are well established and offer a comprehensive collection of software packages. Innovation will, of course, add new tools. It is fascinating to follow the progress of database entries originating from ongoing genome projects, which the reader will discover by comparing some of the compiled data in this book with its status in the databases. This will be proof of the rapid progress in bioinformatics, genomics, and proteomics.

Lukas Buehler
La Jolla, California

Hooman Rashidi
San Antonio, Texas

June 1999

Acknowledgments

Special thanks to Kristen S. Farwell, Becky McEldowney, and Dr. Elzbieta Izbicka for their insightful remarks. Their help is greatly appreciated.

Hooman Rashidi

I owe much of the motivation and discipline to write this book to Lidia Szczupak with whom I share a common interest in the philosophy and history of science, and the belief in communicating one's own scientific work and interests to a general audience.

Lukas Buehler

Authors

Hooman Rashidi, M.S. has been a biological science instructor, researcher, and bioinformatics consultant. He is the founder of The STD Prevention Program in San Diego which he plans to expand nationally. Hooman is a graduate of the University of California at San Diego, where he received a degree with highest honors and departmental distinction in biochemistry and cellular biology. He is also a member of Phi Beta Kappa. His graduate training included the study of computational modeling and phylogenetic analysis of various genes in *Dictyostelium discodium*. He is currently working on computational modeling projects instrumental in rational drug design.

Lukas K. Buehler, Ph.D. has a degree in biochemistry from the University of Basel, Switzerland. His scientific work deals with the structure–function relationship of ion channel proteins found in diverse organisms — from bacteria to humans. His academic research experience includes microbiology, neuroscience, cell biology, and material science. He has also worked on drug discovery projects for business, developing high throughput screening assays using novel genes discovered based on the sequence information originating from ongoing human genome projects. He is the founder of whatislife.com, an Internet forum for the advancement of scientific literacy and is a lecturer in the biology department at the University of California at San Diego.

1

Introduction to Bioinformatics

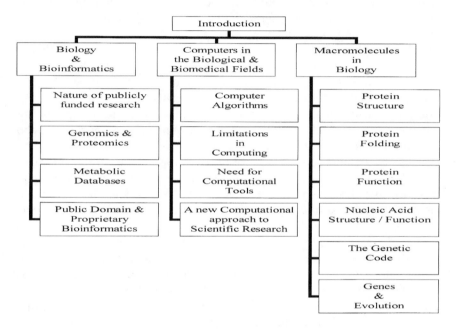

FIGURE 1.1
Chapter overview

1.1 Biology and Bioinformatics

Bioinformatics is a fast-growing field within the biological sciences that was developed because of the need to handle large amounts of genetic and biochemical data. This data, originating from individual research efforts, is linked by its common origin: the cells of living organisms. To understand the links between pieces of information from research areas such as molecular biology, structural biochemistry, enzymology, cell biology, physiology, and pathology, bioinformatics uses computational power to catalog, organize, and structure these pieces into biologically meaningful entities. These "entities"

1

are reflections of the cellular organization of life and its common denominator — that all life evolved from a common ancestral form.

Because bioniformatics is still a young science, everyone defines it differently. It is commonly referred to as the task of *organizing and analyzing* increasingly complex data resulting from *modern molecular and biochemical* techniques. For some, bioinformatics also includes the concept of the flow of information within biological systems alluding to the information stored in the nucleotide sequence — DNA, and its translation into molecules of life — proteins. Bioinformatics is a serious attempt to understand what it means when we say that genes code for physiological traits, like intelligence, brown hair, or susceptibility to cancer. Bioinformatics is presented here as the science of creating and managing biological databases to keep track of, and eventually simulate, the complexity of living organisms. It is based on the assumption that a hierarchical relationship exists among the structure of genes, their arrangement within the genome, the function of proteins, and protein-protein interactions within an organism resulting in energy metabolism, reproduction, and form.

Genes are the hereditary units of life. Therefore, the accurate copying of molecular information contained in genes is crucial for the viability of organisms. However, residual mutation rates are equally necessary for their evolution and the richness of gene pools in populations. Here is a scenario: when scientists publish a report about mutations in the gene involved in the development of breast or ovarian cancers (BRCA-1 and -2), the cloning of this gene is a central first step in obtaining the sequence information and its regulatory components. During this process, a DNA fragment containing the gene must be isolated from the affected individual, isolated from the genome (for BRCA-1 and -2 on the long arm fragment of human chromosomes 13 and 17), and inserted into a small, functionally customized *vector* DNA resulting in *recombinant* DNA. Such vectors allow manipulation of the sequence and the amplification of the gene in a mammalian (other than human) or bacterial cell system making the human gene amenable to genetic manipulation. Once it is an integral part of vector DNA, a human gene can be multiplied in a bacterial cell simply for the purpose of obtaining enough identical copies or clones.

It is important at this point to understand this recombinant DNA technology: the manipulation of human genes in bacterial cells is possible because of the close genetic relationship of all life, a direct consequence of biological evolution as outlined by Darwin's theory of evolution.

Amplifying human DNA in a bacterial cell culture is not only simple and efficient, but also avoids the impracticality of obtaining large quantities of human tissue for the purpose of DNA or protein purification. Successful cloning is followed by the sequencing of the gene to confirm that the intact gene has been isolated and correctly recombined into the vector DNA. The gene sequence reveals a great deal of the expected properties of its corresponding protein structure and function. The recombinant DNA can then be used to synthesize large amounts of protein for biochemical analysis that confirms the predicted structure and function.

In the search for genes of inheritable diseases, samples are needed not only from individuals carrying the mutation, but also from healthy men and women in order to pinpoint the gene loci in the genome (see positional cloning, Chapter 3.1). Indeed, a large sample of afflicted and healthy (control) individuals is needed to establish the suspected correlation between a gene and a disease. Again, the compilation of genetic databases structured according to medical indications enhances our understanding of genetic predispositions for hereditary diseases. Bioinformatics provides the statistical tools for such predictive genome analysis.

Cloning, sequencing, and chromosomal localization, therefore, are three intertwined aspects needed to understand the molecular biology of a gene and its product, the protein. Molecular biology provides the basis for investigating the genotype by means of bioinformatics. Unraveling the relationship between the DNA sequence, chromosomal localization, and structure of genes involved in cancer formation, or for any other inheritable disease, is the true challenge for today's life sciences. Some of the unsolved mysteries in genetics include the relationship between genes and consciousness, genes and heartbeat regulation (so as to avoid arrhythmia), genes and allergies, and genes and behavior — the most profound of all because it forces us to evaluate the essence of being human.

During the early years of the twentieth century, biochemists used organic chemistry to unravel metabolic pathways, study the kinetics of enzymes, and determine the relationships of these pathways to hereditary diseases. This, it should be mentioned, was accomplished without any molecular biology or protein structure information, and some of it was done before it was proven beyond a doubt that nucleic acids — not proteins — are the hereditary units of life. What chemistry did for biochemists in the past, molecular biology is doing for them today. Computer knowledge-based databases will do the same for tomorrow's life scientists.

The Nature of Publicly Funded Research

Not only have techniques changed and influenced the productivity and goals of biochemistry, but the way scientists interact and work in their labs has become a collective approach to solving the truly overwhelming work ahead. Traditionally, understanding metabolic pathways or sequencing genes has been pursued by individual scientists using a clone-by-clone approach. It is often a scientist's personal interest in cancer, heart disease, or memory that spurs research. These individual projects are funded by government agencies such as the National Institute of Health (NIH; http://www.nih.gov/), the National Science Foundation (NSF; http://www.nsf.gov/), hundreds of state agencies, private corporations, and non-profit organizations.

Academic research is powered by individuals who are knowledge driven and whose accomplishments enable them to gain access to grant money. The close relationship between scientists and their accomplishments is fundamental to scientific progress and is closely tied to academic freedom, which involves free access to information while avoiding undue economic pressure. Because of its fast and free flow of information, the Internet has become a vital tool for scientists because of the proliferation and sharing of information contained in centralized databases.

The advantages to scientists of the Internet's information flow are three-fold. First, it provides access to information for anyone with an Internet browser — a form of democratizing information. Second, the data stored in centralized databases shows a varying degree of redundancy. For example, the 12-megabase genome of common baker's yeast, *Saccharomyces cerevisiae*, completed in 1997, has been oversubmitted by a factor of 2.5. Much of this sequence information is the result of yeast being used as a eukaryotic model system in molecular biology. Although this may seem redundant to most people, for biologists, redundancy — as a byproduct of multiple submissions from different authors — provides a *quality control* for both sequence information and biological annotation. Third, when trying to cure human diseases, many genes and biochemical pathways studied in yeast, for example, will help scientists find and understand homologous genes and pathways in humans because of the close relationship of all living organisms based on evolution. This is a new biology of *comparative genomics* across species lines.

The launch in the late 1980s of the Human Genome Project (and many genome projects of other organisms) was a decisive event in the development of bioinformatics. Molecular biologists participating in genome projects are no longer driven by the search for a cellular process and its underlying mechanism and genetic makeup; instead, they simply produce bits and pieces of sequences to be assembled into a complete genomic sequence. Bioinformatics is the analytical tool they need to help obtain this sequence and to unravel the biological information related to these sequences. Genomic information becomes more relevant, because the function of a gene is not only its coding sequence for a protein, but the organization of genes within genomes. Genes are not used in isolation, but in groups, which has long been understood by developmental biologists who track gene activity patterns during the life cycle of an organism. Since every cell in a multicellular organism contains the entire set of genes, but makes use only of a small subset (inactivating many genes that are never used), the combination of active genes in a cell defines its biological fate. This is also known as cell differentiation and is normally an irreversible event during the life of a cell. Functional genomics is a direct consequence of the accumulated information on genome sequences and gene organization within genomes. By analyzing gene activity patterns in cells, tissues, and organs, medical questions are now being addressed, but the technical difficulties and demands of this kind of analysis make it extremely expensive. Because of the potential findings for new drugs, commercial interests are strongly tied to these genome projects and fights over patent rights for yet-unknown DNA sequences have been compared to the gold rush of the nineteenth century.

TABLE 1.1

Reported Data from Species with Genome Sequencing Projects

Species	Major Contributors	Total # of Bases Sequenced	Total Size of the Genome (Bases)
Human	1. Sanger Centre 2. Washington University Genome Sequencing Center 3. Whitehead Institute for Biomedical Research (MIT Center for Genome Research) 4. The Institute for Genomic Research 5. Lawrence Berkeley National Laboratory Human Genome Center 6. Baylor College of Medicine Human Genome Sequencing Center 7. University of Tokyo 8. German Consortium 9. University of Oklahoma 10. Lawrence Livermore National Laboratory Human Genome Center 11. University of Washington 12. Keio University 13. Stanford Human Genome Center 14. Los Alamos National Laboratory	≈140,000,000	≈3,000,000,000
Mouse	1. Whitehead Institute for Biomedical Research (MIT Center for Genome Research) 2. University of Washington 3. Baylor College of Medicine Human Genome Sequencing Center 4. Sanger Centre	≈3,000,000	≈3,000,000,000
Drosophila melanogaster	1. Lawrence Berkeley National Laboratory Human Genome Center 2. Sanger Centre	≈9,000,000	≈165,000,000
Caenorhabditis elegans	1. Washington University Genome Sequencing Center 2. Sanger Centre	Complete	≈100,000,000
Arabidopsis thaliana	1. The Institute for Genomic Research 2. Cold Spring Harbor Laboratory 3. Washington University Genome Sequencing Center 4. University of Pennsylvania	≈7,000,000	≈100,000,000
Schizosaccharomyces pombe	1. Sanger Centre 2. Washington University Genome Sequencing Center 3. Others	Complete	12,067,280
Mycobacterium tuberculosis	1. Sanger Centre	Complete	4,411,525

Genomics and Proteomics

Sequencing efforts by molecular biologists at research universities with interests in diverse aspects of cell biology and biochemistry result in a random collection of gene sequences in various public databases. The advantage of this gene-by-gene approach is in the information about the function associated with the gene. By contrast, the massive approach of sequencing entire genomes leads to the systematic accumulation of DNA sequences without any knowledge about physiology or function, and turns molecular biology upside down. Scientists no longer need to physically screen cell lines or animal tissues to identify a novel gene. Instead, they can electronically screen the public databases for novel, putative gene fragments (electronic Northern) that are the results of high-throughput screening efforts. This provides a shortcut to traditional methods where the isolation of mRNA or sequencing of protein was necessary to get the sequence of a gene. The systematic detection and annotation of protein levels analyzed by 2-D-gel electrophoresis (proteomics; see Chapter 4.1) adds another layer of physiologically relevant information for the detection of novel and important proteins associated with development, aging, and disease. To understand how the genetic blueprint is read and implemented in the development and functioning of a viable organism, this information is fundamentally important.

The promises of modern genome sequencing projects, therefore, are twofold. First, because genomic structures are physiologically relevant, most scientists hope that studying the finished work — the entire genomic sequence information of an organism — will contribute to their understanding of the organism's biology. Second, higher organisms contain massive amounts of non-coding DNA, whose function and existence was questioned and misunderstood until very recently. Having this sequence information, which is currently not accessible through functional approaches, will allow future experimental design to further its discoveries. The promise of genome projects, then, is to understand life. Eureka!

Metabolic Databases

We must then ask: is DNA the only form of information that is inherited? As it turns out, it is not easy to assign inherited information to nucleic acids alone, because DNA cannot replicate itself outside a cellular environment. DNA is a blueprint and has to be "read" somehow. In cells, this "reading" is done by proteins. *Maternal factors,* e.g., pre-programmed RNA, and nuclear and cytoplasmic proteins and enzymes of the female gamete are the tools used to read the blueprint of a fertilized egg and play a crucial role in the early steps of embryogenesis. An effect possibly related to maternal factors is *genetic imprinting,* a control mechanism for gene expression that influences whether the paternal or maternal gene copy is active in the offspring. Genes, which always exist as doubles, get expressed only on one chromosome, inherited from either the father or mother. Chromosomal organization — the interaction and structural arrangement of protein-DNA complexes — is also part of the inherited information and not just the DNA sequence.

To understand phenomena like genetic imprinting and other gene dosis effects (how much protein a gene is able to make relative to a related gene), one has to study spatial and temporal control of cellular processes. The possibilities are tremendous. Once completed, the genome projects will have transformed molecular biology labs into electronic ones (a noble exaggeration), where genetic information will be freely available and will be used to determine the roles of genes in such areas as development, aging, structure and function of proteins, and memory.

Databases are rapidly expanding catalogs, some of which are updated daily to accommodate newly submitted data and make them accessible to the scientific community. By April 1998, the genome projects covered 83 species, with 21 projects completed (predominantly microorganisms) and 62 in progress. These genome projects are powered by automated cloning and polymerase chain reaction (PCR) for DNA amplification and sequencing systems. They are supported by software to reconstruct gap-free, contiguous sequences (contigs) of randomly generated chromosomal fragments (shotgun approach) which eventually leads to the base-by-base sequence of entire genomes. The fragments to be sequenced are no longer chosen based on their known biological significance. Instead, the goal of the genome projects is to "shoot first" and ask questions later — questions that are relevant to obtaining insight into the large unknown portion of life's blueprint. To discover the functions of yet-unknown genes is one of the purposes of most genome projects. The largest project — the Human Genome Project — will not be complete until the last base of the estimated three billion base pairs of the human genome are sequenced and their locations on the chromosomes are determined.

In the meantime, there is a tremendous amount of information available daily from the ongoing genome projects. The usefulness and handling of this information by means of statistical analysis is the topic of this book.

Public and Proprietary Bioinformatics

From the beginning, bioinformatics has been a collaborative approach among different research groups, often from different countries. Bioinformatics is being transformed into an independent science with the advent of centralized data banks and Internet communication, and is fueled by the human genome initiatives which are creating increasingly more powerful cloning and sequencing techniques. Science at its best is an international affair and, consequently, many organizations — private and public — have been founded with the goal of sequencing entire genomes, mapping all genes, and creating relational databases that connect sequence information to functional and structural information at the cellular level.

Industry, realizing the potential of information handling and selling, plays an increasingly dominant role, forcing members of the academic community

Bioinformatics Basics

TABLE 1.2

Progress Report of Databases Frequenly Used in Today's Biomedical and Biochemical Communities

Database	Type of Database	Number of Reported Entries	Reported As Of
SWISS-PROT	Sequence	≈70,000	04-1998
SWISSNEW	Sequence	≈11,000	06-1998
PIR	Sequence	≈110,000	05-1998
EMBL	Sequence	≈2,125,225	04-1998
EMBLNEW	Sequence	≈275,000	06-1998
NRL3-D	Sequence	≈11,000	04-1998
SPTREMBL	Sequence	≈141,000	04-1998
REMTREMBL	Sequence	≈26,000	04-1998
TREMBLNEW	Sequence	≈49,000	06-1998
IMGT	Sequence	≈24,000	04-1998
PROSITE	Motifs	≈1335	03-1998
BLOCKS	Motifs	≈3845	04-1998
TAXONOMY	Sequence Homology	≈59,000	06-1998
PDB	3-D Structures	≈7750	06-1998
OMIMALLELE	Mutations	≈4700	03-1998
OMIM	Mutations	≈8300	04-1997

into a competitive race to become the first to sequence the human genome. Industry, however, uses public databases to tailor its own research, without which proprietary databases would lag far behind in their information. Lines between academia and industry have become blurred and people are moving from one side to the other. The Institute for Genomic Research (TIGR), founded in 1992 by Craig Venter[1], is a premier, non-profit research institute that has grown with and made use of the Internet from the very beginning. For many scientists, TIGR symbolizes the understanding of genome projects and has given new meaning to *high throughput sequencing.* TIGR has been instrumental in developing automated sequence procedures and data analyses. The importance of the service to the public by privately held institutes like TIGR cannot be overstated. The information produced at TIGR and other organizations that can be accessed over the Internet is truly astounding.

The Institute for Genomic Research

TIGR is a not-for-profit research institute with interests in structural, functional, and comparative analysis of genomes and gene products in viruses, eubacteria, pathogenic bacteria, archaea, and eukaryotes (both plant and animal), including humans. The Institute is located in Rockville, MD, in the greater Washington, D.C. metropolitan area, and is close to the National Institutes of Health, Johns Hopkins University, The University of Maryland, and other research institutes and biotechnology companies. Its facilities consist of more than 50,000 sq. ft. of laboratory and office space located on a twelve-acre campus. The Institute has a large DNA sequencing laboratory and modern facilities for bioinformatics, biochemistry, and molecular biology. (from: http://www.tigr.org/about/)

TIGR[1] pioneered the development and distribution of techniques necessary for the mass cloning of DNA fragments and the sequencing of so-called expression sequence tags (ESTs). The hundreds of thousands of electronic sequences deposited in public databases are gold mines for other scientists interested in biologically relevant research.

Many privately held EST libraries (Incyte Pharmaceuticals, Millennium Pharmaceuticals) are being established for medical and pharmaceutical applications and are targeting their products toward commercial buyers. Thus, specific tissue types — diseased or normal, embryonic or adult — build the bases of pharmaceutical and biotechnological proprietary libraries. These cDNA fragment libraries mostly contain sequences of predicted properties, many of which constitute novel genes. Bioinformatics groups in industry subscribe to monthly releases of updated database information for regular screening. Thus, functionally and therapeutically interesting new genes can be identified for drug development.

Scientists are striving to miniaturize the technology for detection of novel genes based on tissue samples rather than electronic databases in order to help in isolating potentially interesting genes. Affymetrix in Santa Clara, California is leading the way to quickly complementing and using the information from genome projects to study activity patterns of genes by measuring the messenger RNA levels on microarray DNA chips[2] containing as many as 15,000 probes on an area of less than a square inch.

References

1. TIGR releases EST data publicly [news]. *Nat. Biotechnol.*, 1997. 15(5): p. 397.
2. Fodor, S.P., et al., Multiplexed biochemical assays with biological chips. *Nature*, 1993. 364(6437): p. 555-6.

1.2   Computers in Biology and Medicine

Applied mathematics and computer science are the tools used in bioinformatics. Today's molecular biology would be impossible without information storage and retrieval, statistical analysis, data fitting, and computer simulation.

Bioinformatics, of course, makes use of computer technology — from PCs using standalone software to the Internet — but computers are also central to almost all other scientific activities. Medical research and treatment, neurobiology, and the use of sophisticated laboratory equipment would be impossible without computers. Modern medicine uses many analytical machines (e.g., biosensors for *in situ* glucose level monitors) and virtual technologies to help physicians in their diagnoses, from inserting miniature probes into tiny blood vessels to performing delicate surgical procedures. Ironically, it is a problem associated with computer timekeeping, the Year 2000 problem, that has greatly increased the public's understanding of our dependency on computer technology.

Neurobiology is beginning to map the brain's anatomy and cellular composition, much as the Human Genome Project maps chromosomes. The brain is a tremendously complex organ and is central to our understanding of what it means to be human. *Neuroinformatics* is a new science emerging from a collaborative effort among neurobiologists, cognitive scientists, and psychologists. The brain — or rather its components, the neurons — is used as a new way to understand complex systems in the form of neuronal networks. Genetic algorithms and non-linear thinking point beyond today's meaning of bioinformatics which links it to artificial intelligence and evolutionary computation (The Genetic Algorithms Archive; an archive maintained by Alan C. Schultz at The Navy Center for Applied Research in Artificial Intelligence http://www.aic.nrl.navy.mil/galist/). A literal meaning of bioinformatics, focusing on the flow of information *within* biological systems, could ultimately be achieved by merging biological molecules into electronic circuits.

Computer Algorithms

Computers are essential in processing large amounts of data in a time-efficient manner. However, computers need instructions — often called "human intervention" — and this human analytical instruction process must be included in estimating the overall time it will take to solve a problem using a computer. Human intervention takes time, so many of today's automated processes have the goal of teaching computers to make decisions in the future (a simple context recognition problem). Expert systems are computer based and are narrowly defined as fulfilling a task that requires enormous computational power. Real-life situations are never totally reproducible and many decisions currently based on human intervention are thought to be able to be handled by neural networks (NNs) that have the ability to learn. Neural networks, although promising, "…are difficult to apply successfully to problems that concern manipulation of symbols and memory. And there are no methods for training NNs that can magically create information that is not contained in the training data" (Warren S. Sarle, Cary, NC, U.S.; from Neural Network FAQ; ftp://ftp.sas.com/pub/neural/FAQ.html). Once an algorithm has been set up and successfully implemented, the computer goes through a repetitive task where input and output data may constantly be changed and adjusted to preset values through feedback and feedforward loops.

The power of computers is undeniable. The ease of writing with a word processor, for example, has made it a very popular and indispensable tool.

Although formatting a text can be done in a matter of seconds, the ease of changing the layout of text and figures has increased the amount of paper waste because we still want to see the tangible end product. The hard-copy version of a text is still more reliable in good copy editing than the electronic version on the monitor. Spell checkers are good examples of simple algorithms that lack analytical measures of the language. One of the pitfalls of spell checking is in proofreading a text within its context — the computer cannot find a typographical error if the typo is a correctly spelled word with another meaning. The analytical ability of the human mind to recognize correct spelling, style, and grammar is not satisfactorily replicated by computers, because computers and minds work differently. Much like proofreading, analysis of scientific data and its subsequent interpretation are aided by computers, but only under the strict control of the human mind.

Computers are excellent tools for numerical solutions (analyses, simulations), controlling and guiding machines, editing information/text, string searches, finding relationships in data, and managing databases. The last three applications are crucial for bioinformatics.

Different Types of Computers for Different Tasks

Personal computers are multifaceted and are used for everything from word processing, spreadsheet analysis, presentation, Internet access, and specialized software to controlling lab equipment such as pClamp from Axon Instrument (http://www.axon.com/), a software application widely used in electrophysiology to control and measure electrical activity of neurons; ion imaging for the measurements of ion and analyte concentration imaging; in genomics projects to analyze the hybridization pattern of DNA chips; and functional neurosurgery for intraoperative microelectrode guidance, and diagnosis and monitoring of movement disorders (e.g., Parkinson's). The versatility, speed, and increased computational power of personal computers and local networks (Pentium processors, NT workstations) allow scientists to do work such as molecular modeling and multiple sequence alignment calculations (molecular evolution) independently of supercomputers. Often, lab equipment comes with optional computer interfaces that allow researchers to customize their applications for specific experimental needs.

It has been estimated that only 1% of all the microprocessors in the world are contained in desktop PCs. The other 99% are embedded in many of the world's products, such as airplanes, heating systems, lab instruments, security systems, and appliances. These processors are often referred to as "firmware," which are chips that have their own special functionality built into them with no programming required.

Science requires the use of many machines with embedded processors, such as gas chromatographs, special computerized scale balances, and spectrophotometers. Spectrophotometers are used to read absorption spectra at different wavelengths, including real-time measurements to monitor changes in the chemical composition of a test solution — the automated fractionation of liquid chromatography separating molecular mixtures into individual components according to their size or solubility properties. These laboratory machines are controlled by microprocessors built directly into the units, rather than by plug-in interfaces and remote computers. The microprocessors are accessible through a small window displaying one or more lines of code or command text that can be typed in or selected from a short menu. They essentially function like ATM machines where a small display and keypad can be used to access your checking account; you are interacting with a computer that allows you to transfer money, but not to type and edit a letter to the bank's customer service department. However, it is connected to a telephone network and is a terminal of a specialized internet. During the last twenty years, these microprocessors have evolved from simple control circuits to "real" computers allowing storage of large data files and graphics, slowly replacing the need for paper recording on plotters and films.

Supercomputers

Supercomputers are used for computationally demanding tasks, and provide appropriate working memory and storage capacity. They are the main hubs of the Internet service providers and, for the most part, use the UNIX operating system. They have also been implemented in many scientific communities, and because of their capacity and price, are prime examples of shared equipment in academia.

The San Diego Super Computer Center (SDSC) is a publicly operated service provider used by many in academia. The SDSC provides and supports a wide range of computing resources. Currently, SDSC's production systems are a CRAY C90 vector supercomputer, a CRAY T3E parallel supercomputer, an Advanced Visualization Laboratory, and an archival storage system. These systems are available for use by academic researchers and students in the U.S. through a peer-review system. They are also available to commercial and government researchers in the U.S. and abroad under special cost-sharing arrangements. Currently, more than 5,100 researchers at more than 240 institutions are using these platforms for their research (source: http://www.sdsc.edu; December 1997).

High speed connections and parallel computing

On June 20, 1997, researchers at the Pittsburgh Supercomputing Center and the University of Stuttgart in Germany linked supercomputers on both sides of the Atlantic via high-speed research networks. This was the first time that high-speed telecommunications networks, such as the very high speed Backbone Network Service (vBNS), had been used for trans-Atlantic metacomputing.

Intended as a prototype for international high-performance networking, the project couples Pittsburgh's 512-processor CRAY T3E with another 512-processor T3E at the High Performance Computing Center in Stuttgart (RUS). Linking two or more supercomputers at different locations to work on the same computing task is known as "metacomputing." The Pittsburgh-Stuttgart link creates a virtual system of 1024 processors with a theoretical peak performance of 675 billion calculations a second. The project relies on a series of research networks to create a high-speed transatlantic link between the two centers. Such networks, established during the past few years, allow information to move up to 100 times faster than on the Internet. For example, the vBNS, which connects U.S. supercomputing centers, currently transmits at speeds of up to 622 million bits per second, fast enough to transfer the complete Encyclopaedia Britannica in less than ten seconds (from *PSC News*; Michael Schneider; Pittsburgh Supercomputing Center; http://www.psc.edu/).

Finally, the Internet is a network of supercomputers and workstations connected by switches, routers, and fiber-optic cables. The great strength of the Internet is its interactive mode. Most interactive tasks available throughout the World Wide Web run through remote supercomputers and are replacing the need to download appropriate software for local analysis, although this is often desirable and applicable with the advent of more powerful PC systems. Bioinformatics is one example of the Internet having become an integral part of research and for which the usefulness of and necessity for standalone programs has become less and less important.

Limitations in Computational Analysis — Paradoxical Promise of the Internet

When an experiment is finished, the accumulated data has to be analyzed or processed. This includes tabulating databases, performing statistical tests for correlations and, most important of all, selecting the data that can be analyzed and used for interpretation. The latter is a process that is independent of the computer and is notorious for relying on the experience of the experimenter. The judgment of the quality of data lies in the eyes of the beholder; i.e., the expected outcome is based on the hypothesis on which the experiment is based.

The intuition and bias of the scientist are the most important factors in making the right decisions. Of course, computers can help, but only with software that has been designed for proper analysis, which means there must be human intervention. While the automation of analytical processes will be done exclusively by computers, the data itself cannot be interpreted by a computer. However, computer-aided data analysis can take out case-by-case biases of the experimenter and read data in a consistent manner that does not reflect the experimenter's biased vision (what the experimenter likes to see).

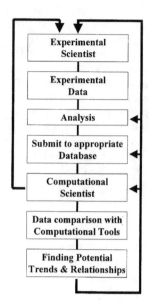

FIGURE 1.2
The marriage between experimental and computational science

No technical manual, disseminated information from books, or the Internet can replace experience. Teaching and gaining experience are the most important factors at this stage of the scientific process. It is here that the Internet offers a seemingly helpful service — that of teaching courses. However, the information found in textbooks, instructional pages, and lab protocols, to name a few, are no replacement for practicing, although the interactive nature of the Internet gives just such a misleading impression. The Internet is actually no more interactive than checking out a book from the library or subscribing to a traditional long-distance college course. In fact, long-distance courses are now being offered by Oxford University through Internet access.

The reason for stressing this point, and understanding that the accuracy of computer-generated data depends on the correct input and handling of data by human operators, lies in the acknowledged unreliability of database annotations. It is no secret that the DNA and protein sequence databases contain errors in the background information of where and how the sequences were obtained. The success of bioinformatics is directly linked to the complete and reliable annotation of database sequence information with accurate biological data. This process of checking annotation for accuracy is not automated and will not be for quite a while. Instead, to make databases reliable sources of information, many specialists in relevant fields are needed to go over the information, word by word.

Comparing sequences of genes using programs like BLAST[1] is fairly easy. However, truly understanding what the results of such a comparison signify can be extremely difficult. The results depend on the types of sequences compared and the origin of the sequences. Both allude to the fact that, besides the mere sequences, one needs to know their functional structure and the means and cellular origins from which they were obtained. In other words, one needs to understand the biology behind the sequence in order to understand the results of a sequence comparison with related or novel sequences. The genome projects produce — and this is, of course, their goal — mostly sequences for which no biological function is associated. The hope is to extract predictive information in order to design experiments to quickly and reliably corroborate the biology behind the sequence.

The Need for Computational Tools

Biology and medicine are now multi-faceted fields concentrating on many aspects of life. Neuroscience concentrates on the biology of the neuron. Biochemistry looks at the chemistry of biological systems. Molecular biology studies biological interactions at the molecular level and their relevance to the cellular and organismic system as a whole. Virology and bacteriology concentrate on viral and bacterial life cycles, respectively. Many other facets of biology concentrate on specific topics that are of relevance to that particular field. While all these facets of biology seem to be distinct, their relevance to one another is indisputable. The overlap among these fields is becoming more obvious with advancements in data management and, therefore, necessitates the use of more elaborate computational tools.

The incredible advancements in biology and medicine during the past several decades have created a positive feedback loop, where each new finding serves as a driving force for the field's growth and popularity. This has also given rise to an exponential growth in biological data. Therefore, the need for an efficient and powerful biological data management system (e.g., NCBI) has become quite obvious, but achieving it without powerful computers is unimaginable. Computers are now an integral part of the biological world and without them, advancements in biology and medicine would undoubtedly be hindered. The partnership between these non-living creatures (computers) and biology has created a need for mergers in certain aspects of the computational and biological sciences. New fields, such as computer science in medicine and computational biology, are on the rise and are rapidly gaining respect in the life sciences. These fields allow for faster analysis of biological data and the discovery of many previously unknown biological trends. These trends, in turn, have been instrumental in therapeutic advancements (e.g., drug design) in prolonging and enhancing life.

The introduction of the Internet to the life sciences has been very rewarding. It has dramatically enhanced communication efforts among the researchers and has minimized repetitive work in various fields. The presence of data management systems such as the National Center for Biotechnology Information

(NCBI) and European Bioinformatics Institute (EBI) have enhanced the efficiency of many research efforts across the world, while uniting life scientists from different disciplines. The exponential expansion of biological data necessitates an organized specification of the data by specialized and specific management systems. For example, the biological data pertaining to proteins should be separate from those of polynucleotides (DNA and RNA). The Protein Data Bank (PDB) library[2] is an example of a data bank where protein data is stored, and deals specifically with protein structures. Like most other biological servers, the PDB server will also provide trends and relationships among the stored molecules. At the PDB, this information could be retrieved from the Structural Classification of Proteins (SCOP).[3] Databases such as SCOP are useful tools in the characterization of the macromolecules with respect to each other and their biological system. The mere separation of the molecules into specific categories is not enough. The data management system should also be able to display relational information regarding the molecule of interest. The information in a particular file should have links to related data on other relevant sites. For example, horse myoglobin's PDB file summary has multiple link options to related information (e.g., related abstracts) for the myoglobin molecule. It is the relational information for a given data entry that makes it valuable to other entries by displaying potential relationships to other molecules and systems in other servers.

New Approaches to Scientific Research with Computers

Unapproachable biological problems are now the subjects of today's world of computerized biology. Biochemistry, molecular biology, evolutionary biology, bioinformatics, neuroscience, and pharmacology are just a few of the fields in the natural science spectrum that have been dramatically influenced by computational tools. In contrast to the physical sciences, the biological fields until recently were considered rather unpredictable and many of their facets were thought to be unrelated. The introduction of computerized tools to the life science field has dramatically reduced its data management problems and has, more importantly, allowed the relationship between the biological molecules and their respective fields to be seen. Biological information and our enhanced predicting ability have strengthened this field and have initiated many collaborations from across its spectrum. The idea of biology as a predictable science is the driving force for many researchers, without which this goal would be dramatically hindered and would seem as if it were science fiction. During the past few decades, due to vast advancements in fields of medicine, the life sciences have received a lot of attention. The introduction of new drugs that could prolong and enhance life has been instrumental in placing many life science fields, including molecular biology and biochemistry, on the pedestal of science.

References

1. Altschul, S.F., et al., Basic local alignment search tool. *J. Mol. Biol.*, 1990. 215(3): p. 403-10.
2. Sussman, J.L., et al., Protein Data Bank (PDB): database of three-dimensional structural information of biological macromolecules. *Acta. Crystallogr. D. Biol. Crystallogr.*, 1998. 54(1 (Pt 6)): p. 1078-84.
3. Barton, G.J., Scop: structural classification of proteins. *Trends Biochem. Sci.*, 1994. 19(12): p. 554-5.

1.3 Biological Macromolecules

Proteins

Amino acids are the building blocks of all proteins. Understanding the fundamental features of proteins requires a profound grasp of their amino acid substituents.

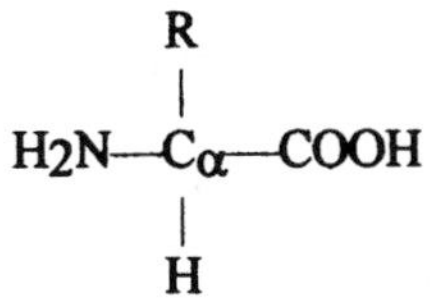

$$R$$
$$|$$
$$H_2N\!-\!C_\alpha\!-\!COOH$$
$$|$$
$$H$$

L - α- amino acid

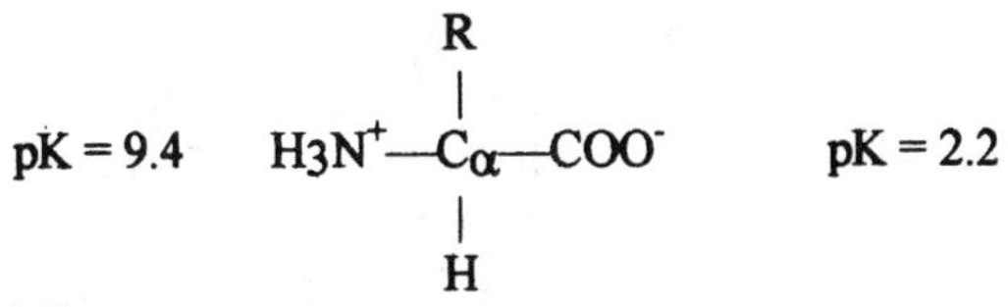

$$pK = 9.4 \quad H_3N^+\!-\!C_\alpha\!-\!COO^- \quad pK = 2.2$$

with the structure:

$$R$$
$$|$$
$$H_3N^+\!-\!C_\alpha\!-\!COO^-$$
$$|$$
$$H$$

Zwitter ionic L-amino acid at physiological pH 7.4

FIGURE 1.3
Chemical structure of an L-α amino acid and its zwitterionic property: an L-α amino acid contains a central alpha-carbon and four chemical substitutents. R represents the side chain with chemical and physical properties outlined in the text. NH2 and COOH represent the basic amino and acid carboxylic group which are always charged at physiological conditions.

Amino acids have both an amino and a carboxylic acid group (Figure 1.3). The amino group adds to their basic nature while the carboxylic acid terminus contributes to their acidity. At physiological pH both termini are charged. This means that the amino terminus stays protonated while the carboxyl terminus sustains its deprotonated form. The distinguishing feature of an alpha amino acid is its residue or side chain. The residues are generally known as the R groups and the characteristic R groups are the distinguishing features of each of the twenty alpha amino acids. Some of these residues are acidic, some are basic, and the remaining residues are relatively neutral.

Acidic residues

Glutamate (E) and aspartate (D):

These are the conjugate forms of the glutamic acid and aspartic acid residues. At physiological pH they are deprotonated and hold a negative charge.

Basic residues

Lysine (K) and arginine (R):

Lysine and arginine are relatively basic at physiological pH. Their basic characteristic promotes their residues to be protonated by their environment and to maintain a positive charge.

Amino acids are also characterized by their relative hydrophobicity. Some are relatively hydrophobic while others tend to prefer polar environments. The relative hydrophobicity of the residue enables us to predict its position within the protein structure. Hydrophobic residues are generally found within the protein core while the polar residues are predominantly found on the surface of the protein structure, interacting with the aqueous environment. The chemistry concept of "like dissolves like" is also applicable to biological systems. Hence, hydrophobic–hydrophobic interactions in most biological systems are preferred to competing hydrophobic-hydrophilic interactions.

- Glutamate, aspartate, lysine, and arginine are charged at physiological pH and are predominantly found on the protein surface, interacting with the polar environment. In general, charged molecules prefer polar environments. This is predominantly due to the polar environment's charge-stabilizing feature (e.g., hydrogen bonding, electrostatic interactions).

- Alanine, valine, leucine, isoleucine, phenylalanine, methionine, glycine, cysteine, and tryptophan are widely accepted as relatively hydrophobic and predominantly found in the protein core and other hydrophobic environments. The presence of carbon chains in these residues adds to their hydrophobic nature.

- Asparagine, glutamine, proline, serine, and threonine residues are the uncharged polar moieties that have a tendency of being solvent exposed.
- Tyrosine's hydroxyl group adds to its hydrophilic nature, while its aromatic side chain contributes to its hydrophobic characteristic. The dual nature of this residue makes it suitable for either environment.
- Histidine is relatively polar. Conformational changes of its ringed side chain contribute to its wide pKa range and adds to its amphoteric nature. Histidine can be either protonated or deprotonated, depending on its environment. This feature makes it a suitable Schiff base in the active site of many enzymes.

TABLE 1.3

General Characteristics of Alpha Amino Acids

Residue	Residue (One Letter Code)	Hydrophobic	Aromatic	Aliphatic	Small	Polar	Charged
Alanine	A	√			√		
Arginine	R					√	√
Asparagine	N				√	√	
Aspartate	D				√	√	√
Cysteine	C	√			√	√	
Glutamate	E					√	√
Glutamine	Q					√	
Glycine	G	√			√	√	
Histidine	H	√	√			√	√
Isoleucine	I	√		√			
Leucine	L	√		√			
Lysine	K	√				√	√
Methionine	M	√					
Phenylalanine	F	√	√				
Proline	P				√		
Serine	S				√	√	
Threonine	T				√	√	
Tryptophan	W	√	√			√	
Tyrosine	Y	√	√			√	
Valine	V	√		√			

√ is representative of the character or partial character of the residues listed

Table 1.4 on the next page is representative of some commonly used hydrophobicity scales. Some are more popular than others, but the important values are those that are consistent in different scales within the table.

The Peptide Bond

Amino acids are linked together via peptide bonds. This is basically an acid-base reaction that results in the loss of a water molecule (Figure 1.4a). In order

Bioinformatics Basics

TABLE 1.4

Some Commonly Used Hydrophobicity Scales

Residue	Residue (One Letter Code)	Kyte/Doolittle [1]	Edelman [2]	Eisenberg [3]
Alanine	A	1.8	0.4397	0.25
Arginine	R	−4.5	−0.7010	−1.80
Asparagine	N	−3.5	−1.414	−0.64
Aspartate	D	−3.5	−2.588	−0.72
Cysteine	C	2.5	1.150	0.04
Glutamate	E	−3.5	−1.270	−0.62
Glutamine	Q	−3.5	−1.656	−0.69
Glycine	G	−0.4	−0.8634	0.16
Histidine	H	−3.2	0.0268	−0.40
Isoleucine	I	4.5	1.546	0.73
Leucine	L	3.8	1.517	0.53
Lysine	K	−3.9	−1.502	−1.10
Methionine	M	1.9	1.746	0.26
Phenylalanine	F	2.8	0.4345	0.61
Proline	P	−1.6	−1.721	−0.07
Serine	S	−0.8	−0.3841	−0.26
Threonine	T	−0.7	−0.0078	−0.18
Tryptophan	W	−0.9	−0.0638	0.37
Tyrosine	Y	−1.3	−0.4585	0.02
Valine	V	4.2	0.5056	0.54

The positive values represent the hydrophobic residues.

to better understand the protein's backbone conformation, one needs to comprehend the nature of the peptide bond and its relationship to the polypeptide's backbone structure. The peptide bond is a special bond that restricts certain angular conformations within the protein (Figure 1.4b). This restriction further limits the total number of possible 3-D conformations for the polypeptide's backbone structure.

Characteristics of the Peptide Bond

The peptide bond has a double-bond characteristic that is due mainly to the resonance system present between its substituents (Figure 1.5). The resonance stabilized nature or double-bond character of the peptide bond contributes to its rigidity and limits its angular rotation. The angular rotation about the peptide bond is also known as omega and is relatively restricted in the polypeptide chain. The double-bond nature of the peptide bond restricts its substituents to a plane and this planarity restricts most of the omega angles to 180 degrees. This is in part due to the cis nature of most di-peptides. In those rare cases where a transconformation is present, the omega angle conforms to a 0-degree conformation. The transconformation is a characteristic of the proline residue. Figure 1.5 displays the partial double character of the peptide bond in one of its resonance forms. The phi and psi angles are also

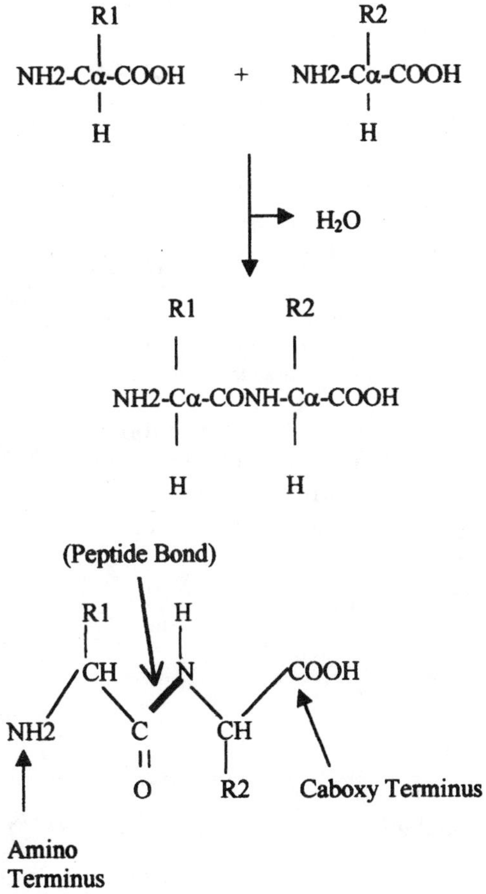

R1   R2
| |
NH2-Cα-COOH   +   NH2-Cα-COOH
| |
H   H

→ H₂O

R1   R2
| |
NH2-Cα-CONH-Cα-COOH
| |

H   H

(Peptide Bond)

R1   H
| |
CH   N   COOH
/   \   /   \   /
NH2   C   CH
|| |
O   R2   Caboxy Terminus

Amino
Terminus

FIGURE 1.4a (top) AND FIGURE 1.4b (bottom)
Formation of the peptide bond: (a, top) Two amino acids covalently link through their amino
and carboxyl group in a condensation reaction releasing one molecule of water and forming a
dipeptide (R1 and R2 linked in the same molecular structure). (b, bottom) The peptide bond
forms a planar structure called the amide plane. The covalent bond between the N–H and C=O
group is not flexible, unlike the C–C and C–N bonds which allow proteins to fold into complex
three-dimensional structures.

marked. Phi and psi are the two main angular conformations associated with
the polypeptide's backbone structure and are used in a two-dimensional plot
to represent a three-dimensional conformation of the protein backbone. This
plot is known as Ramachandran plot and is shown in Color Figure 1.* Phi is
the angular freedom between the alpha carbon and the neighboring N–H
group, while psi is the angle between the C-alpha and the attached carbonyl
substituent.

---
* Color Figure 1 follows page 52.

R1   H   *Phi*

CH   N+   COOH

NH2   C   CH

Psi   O-   R2

FIGURE 1.5
Electronic resonance character of peptide bond: evidence for this structure comes from X-ray crystallographic studies of simple peptides showing that the N–C_α bond length is 1.46Å as expected for a single bond. The C–N peptide bond is 1.33Å long, only a little longer than the value of 1.27Å for the average C=N bond length in model compounds. Similar X-ray studies show the six atoms $C_\alpha NHCOC_\alpha$ as very close to being co-planar.

The protein's linearly extended conformation is also known as its primary structure. Understanding the primary structure of the protein could yield an insight into the molecule's three-dimensional structure.

The following types of interactions drive protein folding:

- Hydrophobic interactions or hydrophobic effect
- Electrostatic interactions
- Hydrogen bonding
- Conformational entropy
- Van der Waals interactions (packing)
- Covalent bonds (e.g., disulfide bridge)

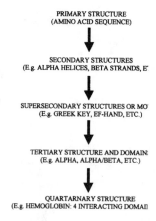

PRIMARY STRUCTURE
(AMINO ACID SEQUENCE)

↓

SECONDARY STRUCTURES
(E.g. ALPHA HELICES, BETA STRANDS, E

↓

SUPERSECONDARY STRUCTURES OR MO
(E.g. GREEK KEY, EF-HAND, ETC.)

↓

TERTIARY STRUCTURE AND DOMAIN:
(E.g. ALPHA, ALPHA/BETA, ETC.)

↓

QUARTARNARY STRUCTURE
(E.g. HEMOGLOBIN: 4 INTERACTING DOMAI

FIGURE 1.6
Protein folding pathway

The Hydrophobic Effect and How It Contributes to Protein Folding

The hydrophobic effect is the preference of hydrophobic or non-polar atoms to stay together and away from the water molecules. This is readily observed when you mix water with oil. The oil molecules are hydrophobic and thus have a tendency to interact with other oil molecules and to reduce their association with the polar water molecules. At room temperature, this effect is predominantly entropically driven.

The hydrophobic effect is widely accepted as the principal driving force in the protein-folding pathway. The environment around most proteins is an aqueous one. Since water molecules are constantly forming and breaking hydrogen bonds with other water molecules, the presence of non-polar side chains in proteins would hinder such a conformational sampling scheme that is entropically favored in the aqueous environment. Hence, the burial of these non-polar side chains is entropically favorable for an energetically sound coexistence between hydrophobic regions of the protein molecule and its aqueous environment.

How do we measure the relative hydrophobicity of a side chain?
This is usually done by using partition experiments. Early experiments by Fauchere and Pliska obtained a hydrophobicity value for each of the side chains by measuring their concentrations through a modeled compound in a medium that represented the protein core and its aqueous environment.[4] The medium they used was octanol. Octanol's long aliphatic chain and terminal hydroxyl group made it a suitable choice for interacting with the non-polar and polar side chains. Some of the hydrophobicity scales and values are presented in Table 1.4 on page 20.

What is the accessible surface area (ASA) and its relationship to hydrophobicity?
The accessible surface area, or ASA, of the solute is defined as the locus of the center of the water probe as it rolls about the surface of the solute.[5]

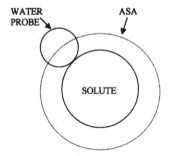

FIGURE 1.7
Solvent-accessible surface area: a solute in water has a specific surface known as the Van der Waals surface. Water molecules will interact with the solute surface and occupy some space which essentially increases the volume of the solubilized molecule. The apparent surface determined by the center of a spherical water moleculae rolled over the Van der Waals surface of the solute determines the water-accessible surface area and represents a useful volume of the solute in modeling drug-receptor interactions.

What is the relationship between the hydrophobicity of the solute and its ASA?
The ASA of non-polar atoms of the extended side chains are to an approximation linearly related to their hydrophobicity.[6] This is not an exact science and the relationship found requires further analysis.

Electrostatic Interactions

This generally refers to the electrostatic interactions present between the ion pairs within the protein molecule. Electrostatic interactions are believed to yield specificity to the protein structure. The effects associated with these interactions are, for the most part, governed by Coulomb's Law. The charge

TABLE 1.5

Non Polar ASA and Hydrophobicity[7]

RESIDUE	Residue (One Letter Code)	Hydrophobicity	Non-Polar ASA (A^{o2})
Glycine	G	−0.06	33
Alanine	A	−0.20	71
Cystein	C	−0.67	98
Valine	V	−0.61	116
Proline	P	−0.44	120
Isoleucine	I	−0.74	140
Leucine	L	−0.65	143
Methionine	M	−0.71	159
Phenylalanine	F	−0.67	164
Tryptophan	W	−0.45	186
Tyrosine	Y	0.22	135
Threonine	T	0.26	76
Histidine	H	−0.04	96
Serine	S	0.34	48
Glutamine	Q	0.74	53
Asparagine	N	0.69	45
Glutamate	E	1.09	61
Aspartate	D	0.72	50
Lysine	K	2.00	118
Arginine	R	1.34	80

of the ion pairs, the distance between the charged groups, and the relative dielectric constants are the key governing features of Coulomb's Law and necessitate maximum interaction between the ion pairs in the protein and its aqueous environment. This is predominantly to maximize a thermodynamically favorable setting for all charged ion pairs. Isolated charges are typically present on the aqueous surface and fully solvated by the water molecules. Disrupting the solvation shells around these isolated charged groups is thermodynamically unfavorable. Therefore, transferring these isolated charged groups to the protein core is less likely to occur. Studies have shown that solvent-exposed charged pairs and isolated groups contribute to the overall stability of the protein molecule. One salt bridge is believed to contribute 0.65 (± 0.35) kcal/mol of energy.[8]

Hydrogen Bonding

It is not quite clear how much hydrogen bonds contribute to the overall stability of the protein structures, but their periodic arrangement within alpha helices and between beta strands provides a plausible arrangement for the formation of secondary structures. Hydrogen bonds are also involved in side chain–side chain and side chain–main chain interactions.[9] The main reason for a doubt in the contribution of buried hydrogen bonds towards protein stability could be attributed to the competition between protein–solvent and protein–protein hydrogen bonds.

Conformational Entropy

The three-dimensional fold of the protein structure limits the sampling scheme of the backbone and side chain (angular) conformations, which is entropically not favored. The Boltzmann equation could be used to estimate the relative loss of conformational entropy when a side chain is restricted to a single rotamer. The restriction of these rotameric conformations is predominantly due to the burial of the side chain within the protein structure.

Van der Waals Interactions (packing)

The folded structure of most proteins favors close packing of its atoms within the molecule. The core atoms are generally more ordered than those that face the surface of the protein molecule. Upon folding, the core of the protein structure is more solid than its surface or unfolded state. This liquid-to-solid transition is an enthalpy-driven event and believed to contribute around 0.6 kcal/mol per CH2 substituent.[10]

Covalent Bonds (e.g., disulfide bridge)

The most significant covalent bond within a protein structure is the disulfide bridge. The disulfide bridge is believed to stabilize the folded structure by restricting certain degrees of freedom of the unfolded chain compared to the same chain without the covalent link. This is entropically driven. Generally, increasing the length of the covalent link corresponds to an increase in stability.[11] This rule is only applicable to a single disulfide bond. The stabilizing effect of the disulfide bond is more profound in smaller proteins.

Protein Functions

Proteins are important biological macromolecules involved in a variety of functions.

Enzymes

These are the biological catalysts. Most of the known enzymes are proteins, without which, life as we know it would cease to exist. Their presence is essential to speeding up biological reactions that are otherwise too slow to sustain life. Enzymes are generally substrate specific and their efficiency is typically dependent on the concentration of the substrate in the cell. This dependency, among others, prevents the enzyme from overproducing products that could overstimulate a response and ultimately cause a catastrophic cellular event. The functional specificity of the enzyme is tightly linked to its structure in three-dimensional space. The three-dimensional conformation of its active site creates its specificity and differentiates one enzyme from another. Therefore, a better understanding of its structural characteristics enables us to have a better grasp of its functional roles. Understanding structural characteristics of proteins with known structures is essential to finding structure–function relationships for sequences whose structures are yet to be determined. A structure–function understanding of these vital proteins could serve as a powerful tool in gaining control over their activities in the event of a malfunction. A malfunction in the activity of these enzymes is believed to be responsible for a variety of pathogenic events.

Regulatory Proteins

These proteins are mainly involved in regulating the activity of other macromolecules within the cell. The concentration of these proteins is responsible for this regulation process. Many of these proteins are involved in a negative feedback regulatory mechanism. In most negative feedback loops, an increase in the concentration of a downstream product hinders the formation of the upstream product. Most feedback regulations are either at the level of transcription (DNA) or at the translation (RNA) level.

Storage

Certain ions, metabolites, or small molecules could be complexed with proteins for the sake of storage. For instance, ferritin stores iron in the liver by complexing the iron ion through its heme group.

Transportation

Certain proteins act as biological transporters. Transferrin and hemoglobin are two separate transporter proteins that carry iron and oxygen, respectively, throughout the body.

Signal

Some proteins are specifically involved in the transmission of biological and cellular signals. Many of these proteins are cellular receptors for small molecules and hormones. Binding of the small molecule or hormone to its respective receptor could cause a signal that is ultimately translated into a cellular response.

Immunity

Most of the macromolecules involved in our immune system are proteins and polypeptides. Immunoglobulins are great examples of a large family of proteins involved in a variety of immunocellular responses.

Structural

A great portion of our proteins have a structural role. These proteins are mainly for mechanical support. Collagen is probably one of the most abundant structural proteins found in all multicellular organisms. It occurs in almost every tissue and is the basis of many cells.

References

1. Kyte, J. and R.F. Doolittle, A simple method for displaying the hydropathic character of a protein. *J. Mol. Biol.*, 1982. 157(1): p. 105-32.
2. Edelman, J., Quadratic minimization of predictors for protein secondary structure. Application to transmembrane alpha-helices. *J. Mol. Biol.*, 1993. 232(1): p. 165-91.
3. Eisenberg, D., et al., Analysis of membrane and surface protein sequences with the hydrophobic moment plot. *J. Mol. Biol.*, 1984. 179(1): p. 125-42.
4. Fauchere, J.L., et al., Amino acid side chain parameters for correlation studies in biology and pharmacology. *Int. J. Pept. Protein Res.*, 1988. 32(4): p. 269-78.
5. Richards, F.M. and T. Richmond, Solvents, interfaces and protein structure. *Ciba Found. Symp.*, 1977. 60: p. 23-45.
6. Chothia, C., Hydrophobic bonding and accessible surface area in proteins. *Nature*, 1974. 248(446): p. 338-9.
7. Miller, S., et al., The accessible surface area and stability of oligomeric proteins. *Nature*, 1987. 328(6133): p. 834-6.
8. Serrano, L., et al., The folding of an enzyme. II. Substructure of barnase and the contribution of different interactions to protein stability. *J. Mol. Biol.*, 1992. 224(3): p. 783-804.
9. Baker, E.N. and R.E. Hubbard, Hydrogen bonding in globular proteins. *Prog. Biophys. Mol. Biol.*, 1984. 44(2): p. 97-179.
10. Nicholls, A., K.A. Sharp, and B. Honig, Protein folding and association: insights from the interfacial and thermodynamic properties of hydrocarbons. *Proteins*, 1991. 11(4): p. 281-96.
11. Harrison, P.M. and M.J. Sternberg, Analysis and classification of disulphide connectivity in proteins. The entropic effect of cross-linkage. *J. Mol. Biol.*, 1994. 244(4): p. 448-63.

DNA and RNA Structure

Nucleic Acid Structure and Function

The molecules central to life can be divided into two groups: nucleic acids and proteins. Because all the bioinformatics that exist today deal with these two classes of molecules, their chemical composition and function in cells will be explained here. Nucleic acids are the hereditary components of life and constitute the genome of every known organism. They are composed of DNA, the deoxy form of ribonucleic acid, or RNA. Proteins are the "tools" used by the cell to read and translate its genomic information into other proteins for performing and controlling cellular processes — metabolism, physiological signaling, energy storage and conversion, and the formation of cellular structures.

Nucleic acids were named after their cellular location when first discovered by Friederich Miescher in 1869. Following Darwin's *The Origin of Species*, published just ten years earlier, the search for hereditary molecules became an important task in chemistry, medicine, and biology. Miescher extracted a substance from nucleic acid that showed acidic behavior and whose solubility depended on the pH of the solution. He called the substance nucleic acid, but did not know that he had isolated the molecular entities of genes. The idea of genes, after all, was not conceived of until long after Gregor Mendel demonstrated that specific traits could be inherited as independent entities (Mendelian view of inheritance of independent segregation and assortment). DNA's function as the sole bearer of genes was demonstrated not only by Oswald Avery, but also independently by Alfred Hershey and Margaret Chase some 75 years after Miescher first described it. Nine years later, in 1953, the double helix structure of DNA was discovered by James Watson and Sir Francis Crick using an X-ray diffraction technique on DNA crystals. The discovery of this structure led to the deciphering of the genetic code. This code determines the information needed for protein biosynthesis and is inherited through a replication mechanism that produces exact copies of the parent genomes to subsequent generations.

The discovery of the genetic code marked the beginning of bioinformatics, because DNA, RNA, and protein sequences could now be identified and established as unique for every type of gene and its corresponding protein. Scientists, however, quickly realized that the simple "one gene–one protein" hypothesis proposed in the early days of molecular biology did not hold true. A complex picture of the gene structure of coding and non-coding DNA emerged, as well as the discovery of recombinant processes that are abundant in all life forms. Molecular biologists learned how to make use of these processes in their laboratory settings, bringing life to recombinant DNA technology — the basis of gentechnology and the proliferation of the biotech industry.

Nucleic acids are linear polymers that are composed of chemically distinct monomers called nucleotides. Nucleotides differ by their composition of aromatic base structure — purines and pyrimidines — linked to a phosphorylated ribose (sugar) unit. Their linear arrangement can be read as sequence of letters designated by a molecular triplet. It is the sequence pattern of these different building blocks arranged in triplets that constitutes the uniqueness of every gene in an organism and of genomes from one organism to another. RNA, the major form of nucleic acid in all organisms, is composed of the ribonucleotides adenosine (A), guanine (G), cytosine (C), and uracil (U), whereas DNA uses the deoxy form of the ribose and the four bases adenine (A), guanine (G), cytosine (C), and thymine (T). DNA, the carrier of genetic information, is found primarily in a dimer or double strand conformation giving rise to the double helix, as shown in Color Figure 2a,* whereas RNA primarily forms monomeric, or single strand units, as shown in Color Figure 2b.* However, RNA can form extended intramolecular double helical domains, as well as hybridize with DNA single strands as shown in Color Figure 2c.*

The double helix structure of DNA is stabilized by electrostatic interactions within pairs of bases from opposing single strands. As a rule, only adenine and

---

* Color Figures 2a, 2b, and 2c follow page 52.

thymine (AT) and guanine and cytosine (GC) form base pairs in chromosomal DNA. The precision of this pairing lies in the thermodynamic and conformational limitation of *hydrogen bond* formation between aromatic ring structures of nucleotides.

The Genetic Code

The genetic code is the key to properly translating a sequence using four different nucleotides into a sequence using twenty different amino acids. A nucleotide triplet defines the position of an amino acid in the protein. Once established, a unique feature of this process was observed in all bacteria, plants, and animals; translation is only possible from nucleic acid to protein and not vice versa. Nucleic acids serve as templates for protein synthesis, but genes cannot be synthesized from proteins. This is the *central dogma of molecular biology* and reflects the way all living, cellular organisms reproduce and grow.

Certain groups of viruses contain RNA molecules instead of DNA in their genome. One would expect, according to the central dogma, that these viruses use their RNA to directly guide protein synthesis in their host cells. Yet it has been found that once inside the host cell, these viruses use a protein that is able to copy the viral RNA genome into a DNA copy (so-called cDNA or complementary DNA). Once inside the host cell, the protein integrase catalyzes the stable insertion of this cDNA into the host genome. The protein responsible for the synthesis of DNA from RNA templates, a process which is not found in any non-viral organisms, is called *reverse transcriptase* and the laboratory use of this protein has revolutionized molecular biology.

Reverse transcriptase, together with heat-stable DNA polymerases from thermophilic bacteria, is the cornerstone of today's molecular biology. Polymerase chain reaction (PCR) and cDNA formation allow for the rapid identification of novel genes even from tiny tissue fragments, a feature prominently exhibited in forensic science. Because of the increasing numbers of DNA sequences (and amino acid sequences) published based on these two complementary techniques, bioinformatics was born out of the need to store, process, and annotate genetic information.

The genetic code is universal and redundant. Knowing this is important in understanding the relationship between the genomic organization of an organism and the structure of its proteins and the diversity of life.

The genetic code is redundant because a total of 64 possible triple-base combinations (codons) can be used for a set of 20 amino acids. Reserving stop or termination codons, several amino acids are coded by more than one codon, including the start codon for the amino acid methionine. This means that the amino acid sequence is more conserved than the DNA sequence. This has implications for the mechanism of evolution because some single-base exchanges do not result in changes in amino acid sequence. These single-point mutations are also called *silent mutations* because they do not affect the phenotype and therefore are not subject to selective pressure. Redundancy, however, is often limited because organisms do not make use of the full set

30

Bioinformatics Basics

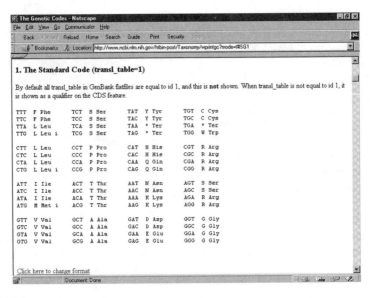

FIGURE 1.8

The genetic codes — standard code: translation table site for the translation table 1 (includes human) at NCBI's taxonomy site. A total of 11 codes have been described in nature, including eukaryotic organelles and the alternative yeast nuclear code.

of codons, but instead use a selection (Figure 1.8). This is known as codon usage or bias. This selection can differ from organism to organism. Codon bias is important for recombinant DNA technology experiments where genes of one organism are cloned and recombined into another organism's genome for manipulation and expression purposes. In some cases, codon bias can result in nonfunctional proteins or affect the level of protein synthesis. Codon bias can also serve as protection against foreign DNA from pathogenic organisms. Unused codons have the effect of stop codons in foreign DNA, effectively inhibiting synthesis of functional proteins necessary for the reproduction of pathogenic organisms.

Codon bias, however, is the result of an otherwise universal use of the genetic code. Except for some organellar DNA, all organisms, including viruses, use the same codons for the 20 amino acids used for protein synthesis. This means that genes can be transferred between organisms, which is the basis for today's biotechnology industry. It also means that bioinformatics does not have to distinguish sequence information based on its cellular origin. This universality increases the sample base for the statistical analysis of DNA sequences stored in databases, and allows well-studied genes from model organisms such as rat, fruit fly, or *Escherichia coli*, to be easily compared with their human counterparts. DNA sequence similarity allows for the rapid identification, cloning, and sequencing of genes in related organisms. In addition, missing biological information in one organism can be inferred, although with caution, from other species. What "works" in a fruit fly or a nematode may well be compared to human metabolism.

Genes and Evolution

Genes are the hereditary units of all life and are made of deoxyribonucleic acid, or DNA, except for some viral genomes such as the human immunodeficiency virus (HIV, a retrovirus) that are made of RNA. Based on a morphological criterion pertinent to their genome, namely the presence or absence of a cell nucleus, all organisms can be grouped into two major groups: the eukaryotes and prokaryotes. The latter are single-cell organisms and are subdivided into two urkingdoms: eubacteria and archaea, respectively.

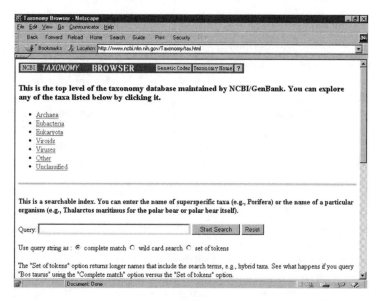

FIGURE 1.9
Taxonomy browser at NCBI. Top level: the main menu of NCBI's taxonomy browser offers seven links to distinct "classes" of organisms. They include three urkingdoms: archaea (single cellular), eubacteria (single cellular), and eukaryota (single and multicellular; contain nucleus and other internal organelles). The other links refer to viruses and unclassified organisms. Viruses are not considered complete life forms and need the cellular machinery of the organisms of the three urkingdoms to proliferate. The taxonomy browser contains only organisms for which an entry exists in one of the databases: nucleic acid sequence, amino acid sequence, or protein structure.

Although having similar morphological characteristics, eubacteria differ in their basic genomic structure from archaea. The grouping into three urkingdoms (the eukaryotes, the eubacteria, and the archaea) has been based mainly on analysis of ribosomal RNA; however, with several complete genomes from members of all three urkingdoms, the archaea branch appears to have eukaryotic, as well as eubacterial characteristics, depending on which set of proteins and metabolic pathways are being studied.

The question of the correct tree of life is as old as taxonomy, and bioinformatics simply adds a new analytical tool for finding the answer to this question. It is clear, however, that neither genotypical nor phenotypical taxonomy alone can solve the problem. Molecular biology does not simply replace "older" outdated branches of evolutionary biology; it is, at best, supplemental. Questions regarding the origin and coexistence of two basic genomic structures have fueled an ongoing debate among evolutionary biologists regarding which one occurred first and how they came into existence. The genome projects involving organisms from all three urkingdoms will certainly contribute to our understanding of this problem.

Eubacteria

Eubacteria are prokaryotic single-cell organisms that contain genomes with a highly compact gene structure and organization. Simply put, all genes contain a single coding region that corresponds to the amino acid sequence of the protein it is coding for. These coding regions are flanked by control regions that define the way proteins have access to the DNA for replication and transcription. Often, genes are grouped within functional units that are regulated in a coordinated manner. These genes code for several enzymes that constitute metabolic pathways. The corresponding multi-gene structures are called *operons*. The operon reflects the functional unit of the gene that is working in concert and its down and up regulation, e.g., gene expression, is coordinated by a single transcriptional unit.

Eukaryotes and Archaea

The genome structure in archaea and eukaryotic organisms is more complex than in eubacteria. Most of their genes are not simple, single-coding frames, but are fragmented into exons and introns which are coding and non-coding regions, respectively. Eukaryotic genomes contain only about 5% to 15% coding regions, or genes, meaning that the vast majority of DNA is either not coding for any proteins or is not known to be doing so. This DNA may be important in the regulation or "behavior" of the genome as a whole, specifically during meiosis, an important process in the reorganization of chromosomal DNA during sexual reproduction. Archaea, which are morphologically indistinguishable from eubacteria, are more closely related to eukaryotes in their genome structure and certain — but not all — metabolic pathways. With access to several completed genomes, it becomes evident, however, that the classification into urkingdoms is strongly dependent on the metabolic group of enzymes studied. The archaea may be the oldest group of organisms — in other words, the modern group of organisms that most closely resembles the suspected common ancestor of all life on earth. Again, it is the universal existence of the genetic code that strongly suggests such a single, common ancestor organism.

2

Databases and Search Tools

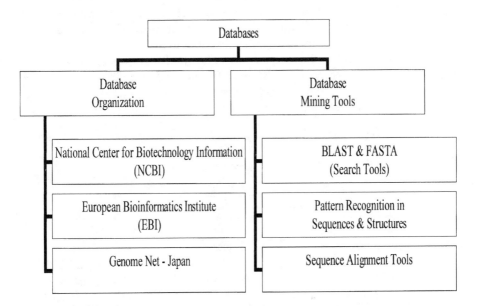

FIGURE 2.1
Chapter overview

2.1 Computational Tools and Databases

Today's life scientist is interested in discovering both the biological code that is inherent in all living things and its significance in fighting pathogens. Solving the triplet code in DNA molecules and its relationship to the translated product was the first major step in gaining insight into the mechanism of most carbon-based life forms. Biology is no longer an empirical study of living things; the exponential growth of biological data in the past few decades has added a predictive element to this field. Today, many biological events can be explained using basic physical and chemical laws. With additional data, many of today's biological ambiguities will eventually be explained. This will facilitate the discovery of new biological trends and laws crucial to our understanding of these complex systems.

Computational tools and databases are essential to the management and identification of subtle patterns found by using this exponentially growing volume of biological data. The National Center for Biotech-nology Information (NCBI) in the United States[1] and the European Bioinformatics Institute (EBI) in England[2] are two main life science servers responsible for dealing with this staggering volume of data. They both maintain reliable databases and analytical software that serve as valuable tools for today's scientific community. New entries to their databases are submitted every day and their busy scientific staff adds the new data to the appropriate database. This allows the scientific community that subscribes to their databases to stay well informed while facilitating the progress of a variety of projects. The services offered by the servers (e.g., NCBI and EBI) are made possible by fast, elaborate computers that can perform the necessary analytical tasks, and the Internet interface that facilitates the electronic communication efforts.

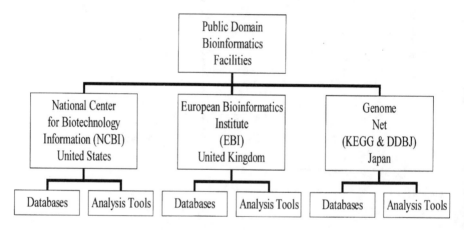

FIGURE 2.2
Primary public domain bioinformatics servers

National Center for Biotechnology Information (NCBI)

In November 1988, the U.S. Senate recognized the need for computerized data processing in the biomedical and biochemical fields and passed legislation that helped to establish NCBI at the National Library of Medicine (NLM). NLM's focus is on maintaining biomedical databases, while NCBI is specifically involved in the development of new analytical tools to aid in understanding the molecular and genetic processes that are key players in pathogenic events. NCBI's four main tasks are:

1. To create automated machines that can analyze and store data pertaining to molecular biology, genetics, and biochemistry;
2. To facilitate usage of the database and analytical software available to the scientific community (e.g., researchers, medical staff, etc.);

3. To coordinate worldwide efforts to gather biological data;
4. To conduct research in computerized analysis of structure–function relationships for key biological molecules.

Who is employed by NCBI?

NCBI's scientific staff consists of computer scientists, molecular biologists, mathematicians, biochemists, research physicians, and structural biologists.

What kind of research is conducted at NCBI?

The collaborative effort of NCBI's staff allows them to study the molecular basis of disease by using mathematical and computational tools. The three main facets of their studies are:

1. To analyze the sequence of the gene or gene product of interest;
2. To gain a better understanding of the organization of the genes analyzed;
3. To predict the structures of the molecules analyzed (e.g., proteins).

The analysis step could include the comparison of novel sequences to known homologs by comparing the sequence of the unknown protein or polynucleotide to proteins or polynucleotides whose sequence is known. Understanding the organization of the genes with respect to the whole genome could be a powerful tool for analyzing future novel genes whose functions are unclear or those that lack homology to any of the known sequences in the database. The step involving structure prediction of structurally unknown molecules, using homologs whose structures are known, would enable us to predict potential functional characteristics associated with molecules whose structure and function are not yet known.

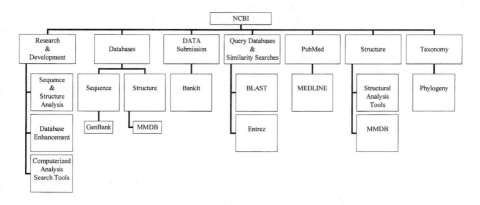

FIGURE 2.3
National Center for Biotechnology Information (NCBI)

What types of databases are supported by NCBI?

- Protein sequence: these are experimentally sequenced proteins and translated nucleotide sequences from nucleotide libraries.

  a. Redundant protein sequence databases (e.g., PIR's complete database[3] which consists of PIR1+PIR2+PIR3)

  b. Non-redundant or less redundant protein sequence databases (e.g., NR, SWISS-PROT,[4] and PDB[5])

- Nucleotide sequences (DNA and RNA): these are DNA and RNA sequences derived from less-automated sequencing projects (e.g., Genbank[6]) or automated sequencers (e.g., dbEST[7]).

  a. Redundant nucleotide sequence databases (e.g., dbEST)

  b. Non-redundant or less-redundant nucleotide sequence databases (e.g., Genbank)

What do we mean by redundancy?

Redundancy in biological databases is a rather complicated issue. Should two alleles from the same locus be considered as one? What about the functionally identical enzymes (isozymes) in the same organism? How about tissue specificity in proteins and their relationship to their respective homologs in other tissues? These are all valid issues and therefore require each database to have its own definition of a redundant sequence. Most databases use automated measures to account for redundancy, especially in large projects. This method is qualitatively less sensitive than manual intervention, but makes up for it in speed. On the other hand, non-redundant databases allow redundant sequences for the sake of completeness.

The following is a list of the most frequently used protein sequence databases at NCBI:

- Alu: This is a selected set of translated alu repeats.[8] This allows the masking of the potential alu repeats in the query sequence. This database can also be retrieved through NCBI by anonymous FTP (under the/pub/jmc/alu directory).

- *E. coli*: This database specifically carries *E. coli* genomic CDS translations.[9]

- Kabat: This database deals with sequences that have immunological interest.[10]

- Month: This is a database of new or recently revised (within the last 30 days) CDS translations from GenBank and other protein sequence entries at the PDB, SWISS-PROT, and PIR libraries, combined.

- NR: This is a database of all non-redundant CDS translations from GenBank and other protein sequence entries at the PDB, SWISS-PROT, and PIR libraries, combined. In this database, the proteins with identical sequences are merged into a single file.

- PDB: these are protein sequences whose 3-D structures are known. The Protein Data Bank (PDB) structures[5] are at the Brookhaven National Laboratory in Long Island, New York. This information is also found in NCBI's PDB mirror site in MMDB. The entries at PDB are predominantly non-redundant. In the case of identical sequence entries with multiple structures, the entry with the highest quality structure is kept. For crystals structures, this is generally the entry with the smallest resolution value (e.g., 1.8 Å preferred over 2.2 Å). However, other variables, such as complexed structures with bound metals or ligands, will allow multiple structures for the same sequence in a given organism.

- SWISS-PROT: this is a database of the most recent release of protein sequence entries from the SWISS-PROT database.[4] It is now supported by EBI, an outstation of EMBL. This is one of the most informative cross-referenced, protein-sequence libraries available through the Internet. SWISS-PROT is a non-redundant database maintained by Amos Bairoch at the University of Geneva.

- Yeast: the yeast (*S. cerevisiae*) protein sequence database[11] stores the sequences generated from the yeast protein sequence projects.

The following is the list of some of the most commonly used nucleotide databases available through NCBI:

- alu: This allows the masking of the potential alu repeats in the query sequence. This database can also be retrieved through NCBI by anonymous FTP (under the/pub/jmc/alu directory).

- dbEST: this is a questionably non-redundant database of GenBank, EMBL, and DDBJ EST entries. ESTs are single-pass cDNA sequences generated through automated sequencers with little or no human intervention. This will therefore increase the error frequency observed in these sequences relative to the rest of the sequence libraries. The most common errors observed in these entries are sequencing errors, heterologous sequence contaminations, and the presence of transcribed repetitive elements.

- dbSTS: this is a non-redundant database of GenBank, EMBL, and DDBJ STS entries.

- E. coli: this database specifically carries E. coli genomic nucleotide sequences.

- EPD: eukaryotic promotor database[12] contains a list of all known eukarotic promoter sequences in public domain libraries.

- GSS: the genome survey sequence contains single-pass genomic data, exon-trapped sequences, and alu PCR sequences.

- HTGS: This is the high-throughput genomic sequences database.

- Kabat: this database deals with sequences that have immunological interest.
- Mito: this database specifically deals with mitochondrial sequences.
- Month: this is a database of new or recently revised (within the last 30 days) entries that are found in GenBank+EMBL+DDBJ+PDB sequence libraries.
- NR: this is a database of all non-redundant GenBank+EMBL+ DDBJ+PDB sequence entries. This database excludes EST, STS, GSS, or HTGS sequence entries. Entries with a 100% sequence identity are merged as one.
- PDB: these sequences are derived from the three-dimensional structure of the molecule.
- Vector: This is the vector subset of GenBank (NCBI's nucleotide sequence database).
- Yeast: the yeast (*S. cerevisiae*) genomic nucleotide sequence database stores the sequences generated from the yeast genome project and other relevant yeast sequencing projects.

What are some of the services offered by NCBI?

The following are the seven main databases and analysis tools supported by the NCBI server at their web site:

1. PubMed (Public MEDLINE)
2. BLAST: Basic Local Alignment Search Tool[13]
3. Entrez[14]
4. BankIt (World Wide Web submission)
5. OMIM (Online Mendelian Inheritance in Man)[15]
6. Taxonomy
7. Structure

Pubmed

PubMed is the search service of the National Library of Medicine (NLM). It allows the user to gain access to over 9 million citations in MEDLINE and pre-MEDLINE, and is linked to participating online journals and related databases enabling the user to retrieve pertinent information in a speedy and efficient manner. Keywords may be used to retrieve journal articles that contain relevant topics. Multiple keywords may be used to increase the specificity of the search. Other search criteria such as author names and journal titles are also available for the user's convenience.

BLAST: Basic Local Alignment Search Tool

The Basic Local Alignment Search Tool[13] is a set of similarity search programs that are designed to identify the classification and potential homologs for a given sequence. These programs are robust and capable of analyzing both DNA and protein sequences. BLAST programs are explained in further detail in Chapter 3.2.a.

Entrez

The scientific researcher is obligated to produce original non-redundant data that will serve to enhance the understanding of a particular principle. To prevent or minimize redundancy in published material, scientists must ensure the originality of their findings. This is not an easy task, but elaborate search tools with accessibility to relevant databases can facilitate the process. For instance, if a researcher has identified a particular trend in a family of proteins, the obvious next step would be to ensure the originality of his work. In other words, is this a new finding? To answer this not-so-trivial question, all possible citations with similar keywords would need to be searched. The results of this search could follow three main paths: the first path could lead you to a redundant dead end. In this situation, the data would be identical or very similar to previously answered questions. At this point, wise researchers would stop working on the redundant data and try to refocus their energy on other findings. The second path could lead to an original end point with no similar findings. In this case, the findings would be completely unrelated to anything in known citations. This could be good or bad; it could either signify an original finding or a mistake. At this point, the researcher would need to further investigate the findings and verify the steps followed in the protocol. This could further support the findings or allow the investigator to find the potential errors in either the protocol or the data analysis steps. The third path could lead to a relevant end point where citations would be found supporting the recent data without restating the same finding. This is the ideal situation for an investigator. The relevant citations could then be used as supporting references for the new findings. In any case, to conduct a reliable search a researcher must utilize a search engine that is not only efficient, but has access to all relevant, regularly updated databases. The government-supported search tools are generally the most reliable software available in the public domain. They are readily accessible through the Web and are very user friendly.

One of the most popular search engines is the Entrez at NCBI.[14] The Entrez Web interface (http://www.ncbi.nlm.nih.gov) allows the subscriber to gain access to bibliographic citations and biological data from a variety of reliable databases. Protein sequence information is retrieved from SWISS-PROT, PDB, PIR, and PRF. Proteins whose structures are known are retrieved from Brookhaven PDB. These proteins are incorporated into NCBI's Molecular Modeling Database, also known as MMDB.[16] The translated proteins and

DNA sequences are retrieved from their parent DNA sequence databases (e.g., GenBank, EMBL, and DDBJ). For a bibliographic or citation search, Entrez uses PubMed's bibliographic database which has access to over nine million biomedical articles from MEDLINE and pre-MEDLINE databases. Entrez also has access to chromosome mapping and genomic data. (PubMed's MEDLINE searches figures, gif files). Entrez offers a variety of criteria for a particular search. For example, one could search a relevant database to find all possible terms that begin with a given word. Placing an asterisk at the end of the term allows Entrez to search for all possible words that begin with that particular term. For example a search for "inter*" will find all terms beginning with "inter," such as interstetium, intermolecular, etc. Entrez can also be used to conduct a smart search during which Entrez will search for phrases or groups of words. Entrez will automatically group the relevant terms together and exclude the unrelated terms from the grouping.

For instance, to locate all the possible citations from a particular author (e.g., Rashidi HR) that deal with a given subject (e.g., energetics), the user can enter the individual terms known about the author (e.g., Rashidi HR) and the subject of interest (e.g., energetics). Entrez will automatically recognize and group the relevant terms (e.g., the author's last name and initials), allowing the search engine to seek all relevant material from Rashidi HR that deals with energetics ("Rashidi HR" AND energetics). Entrez can also be made to group words that otherwise would be considered separate terms by using the automated grouping task. Inserting quotes would cause Entrez to group seemingly irrelevant terms into one (e.g., "brca 1"). Nevertheless, NBCI recommends that users allow Entrez to group the specified terms to minimize inaccurate retrievals. If the retrieval list from the search result is too long, Entrez will halt the search operation and inform the user.

One of the most accurate ways to retrieve a particular citation or sequence is through its identifier. An identifier is an index number assigned to a particular sequence or article in its relevant database. For instance, the identifier for MEDLINE citations is referred to as an UID number, while identifiers that pertain to a sequence are called GI numbers. To retrieve a MEDLINE citation with the UID 88067898, the user would simply input "UID 88067898" in the Entrez search engine to find the MEDLINE citation that is assigned this UID.

There are numerous search files on Entrez, and experienced users find it to be useful and time efficient due to its adaptable nature. The following are some of the search fields that can be customized to meet the user's specific needs:

- **Keyword** allows the user to search a set of indexed terms associated with NCBI's accessible databases (e.g., GenBank, EMBL, PDB, DDBJ, SWISS-PROT, PIR, or PRF).
- **Accession** allows the user to search accession numbers assigned to proteins, nucleotide sequences, structures, or genomic records.

- **Author Name** has information about the authors of published papers. These are typically MEDLINE articles.
- **Affiliation** is used to search for the author's institutional affiliation and address.
- **Journal Title** is used to search for the name of the journal where the record was published. The user may utilize the **List Terms** mode to browse the list of abbreviated journal names (e.g., the *Journal of Biological Chemistry* is abbreviated as "J Biol Chem").
- **E. C. Number** is a designation number assigned to enzymes by the Enzyme Commission.
- **Feature Key** can be used to search keywords denoting a particular DNA feature.
- **Gene Symbol** can be used to search standard names for given genes.
- **MEDLINE UID** is used to search citations using a MEDLINE identifier.
- **MeSH Terms** are used to search Medical Subject Headings. These are a set of keywords used to index MEDLINE.
- **MeSH Major Topic** includes all the terms in MeSH tagged by the indexers as being of major importance.
- **Publication Date** is used to search for the date the article or sequence was published or submitted.
- **Modification Date** is the date the record was placed into Entrez.
- **Page Number** contains the page number of the published article.
- **Property** tells the user what type of sequence the citation contains.
- **PubMed ID** is PubMed's identifier for a given citation.
- **Organism** is used to search for the common and scientific names of the organisms associated with the protein or nucleotide sequence entries.
- **Protein Name** is used to search for the name of the protein a sequence is associated with.
- **SeqId** is a string identifier for a given sequence.
- **Substance** is used to search for the names of chemicals associated with the records from Chemical Abstract Service (CAS) registry.
- **Title Words** is used to search for words that are only found in the title line of a record.
- **Text Words** is used to search for "free text" associated with a given record. For protein and nucleotide sequence records, this includes the definition, comment, name, and description of the given

sequence. For MEDLINE entries, this includes the title and the abstract of the given record.

- **Volume** is used to search for the number of the journal volume that contains the article of interest.

If the specified search field does not find the records of interest, it is helpful to repeat the search using "All Fields" or "Text Words." The intersection symbol is translated as AND in Entrez, and will only seek records that contain all of the given terms separated by AND or AND symbols. Entrez recognizes the union symbol as OR, which allows the user to find documents that contain any of the given terms. Finally, the difference, or BUTNOT, option enables the user to find all the documents that contain the uppermost terms, but not the lower terms.

After a successful search, the user is given retrieval options in the list of documents meeting the given criteria. The list of search results appears in chronological order from the most recent records to the oldest relevant records on file. The user can either retrieve all the documents or select the most relevant reports from the list of records found. The following are several different viewing formats for the relevant retrieved files: PubMed articles can be viewed as Citation, Abstract, MEDLINE, or ASN.1 type formats. Citation formats display the title, abstract, MeSH terms, and the substance information of an article. The Abstract formats display only the title and the abstract of the article. ASN.1 is a special format used by PubMed articles, while MEDLINE displays the article in MEDLARS format. GenBank/GenPept, Report, ASN.1, Graphic view, and FASTA are some of the viewing format options for protein and nucleotide records. GenBank/GenPept format is the standard GenBank or GenPept database file. Report allows the user to view the sequence record as a GenBank report format. Graphic View enables the user to display a graphical view of the sequence entry including alignment information. The FASTA format is most useful for further analysis of the given entry.

Many of the alignment tools (e.g., BLAST) require the user to input the order of interest in a FASTA format. The viewing options for structural information are the Structure Summary and the ASN.1 formats. The Structure Summary format is used to gain access to a summary of the structural data available for a given molecule. For instance, in crystallized protein structures, this view allows the user to gain access to information regarding the resolution of the given structure, author information, date of submission, complexing ligands, and other basic information. This format also allows the user to view the 3-D representation of the molecule. The graphical view is also an option for genomic records.

All the formats described can be saved as documents in a user's file. The three primary save options are: Text, HTML, and MIME. The MIME format is of particular use if the user has access to GenBank's MIME viewer. Otherwise, the output file must be saved in a text or HTML format to be of use. The HTML format is useful if the results will be viewed through a Web browser.

The Text format lacks the HTML tags and breaks but can be viewed with standard word processing software such as Microsoft Word.

BankIt

BankIt is GenBank's World Wide Web sequence submission server. It allows the user to submit new sequences to GenBank via a user-friendly Web browser. The sequence and all relevant information are pasted into a submission box and sent to GenBank. GenBank's staff then contacts the submitting party and assigns an accession number to the sequence.

OMIM (Online Mendelian Inheritance in Man)

This database of human genes and gene disorders is maintained by Dr. Victor A. McKusick, his colleagues at Johns Hopkins University, and other contributors.[15] The OMIM Morbid Map is also supported on this site and maps genetic locations based on and organized by genetic disorders. Entrez, GDB, the Davis Human/Mouse Homology Map, the Online Mendelian Inheritance in Animals (OMIA), the Human Gene Mutation Database (HGMD), the Alliance of Genetic Support Groups, the Cedars-Sinai Medical Center Genetics Image Archive, the Jackson Laboratory, RetNet (retinal genetic disorders), HUM-MOLGEN, and the locus-specific mutation databases are some of the resources available at OMIM. This site is typically used by physicians and medical investigators concerned with genetic disorders. Having a solid understanding of scientific concepts and procedures is necessary for optimal interpretation of the images and text found at OMIM.

Taxonomy

NCBI's taxonomy home page contains organismic databases with the scientific and common names of organisms for which some sequence information is known. The server allows the user to access species' genetic information and how it ties in with related and not-so-related species. Trees are typically a representation of such relationships. These relationships can be based on similar proteins or nucleotide sequences. This page also has links to other NCBI servers (e.g., Structure and PubMed).

Structure

The Structure home page at NCBI supports the Molecular Modeling Database (MMDB) and a variety of software tools relevant to structural analysis.[16] The MMDB information is obtained from the Brookhaven Protein Data Bank (PDB). This includes the X-ray crystallography or Nuclear Magnetic Resonance (NMR) determined structures of important biological macromolecules. Cn3-D[17] is NCBI's structural visualization software for MMDB and is available at the Entrez/Cn3-D FTP site. Structure also offers research tools such as PKB and Threading. This software is available through the FTP site and requires Splus. The site's Entrez/PubMed link facilitates the search for applicable and related information to the molecules of interest.

References

1. Woodsmall, R.M. and D.A. Benson, Information resources at the National Center for Biotechnology Information. *Bull. Med. Libr. Assoc.*, 1993. 81(3): p. 282-4.
2. Emmert, D.B., et al., The European Bioinformatics Institute (EBI) databases. *Nucleic Acids Res.*, 1994. 22(17): p. 3445-9.
3. Barker, W.C., et al., The PIR — International Protein Sequence Database. *Nucleic Acids Res.*, 1998. 26(1): p. 27-32.
4. Bairoch, A. and R. Apweiler, The SWISS-PROT protein sequence data bank and its supplement TrEMBL in 1998. *Nucleic Acids Res.*, 1998. 26(1): p. 38-42.
5. Sussman, J.L., et al., Protein Data Bank (PDB): database of three-dimensional structural information of biological macromolecules. *Acta. Crystallogr. D. Biol. Crystallogr.*, 1998. 54(1 (Pt 6)): p. 1078-84.
6. Benson, D., D.J. Lipman, and J. Ostell, GenBank. *Nucleic Acids Res.*, 1993. 21(13): p. 2963-5.
7. Boguski, M.S., T.M. Lowe, and C.M. Tolstoshev, dbEST — database for "expressed sequence tags" [letter]. *Nat. Genet.*, 1993. 4(4): p. 332-3.
8. Moyzis, R.K., et al., The distribution of interspersed repetitive DNA sequences in the human genome. *Genomics*, 1989. 4(3): p. 273-89.
9. VanBogelen, R.A., et al., The gene-protein database of *Escherichia coli*: edition 5. *Electrophoresis*, 1992. 13(12): p. 1014-54.
10. Martin, A.C., Accessing the Kabat antibody sequence database by computer. *Proteins*, 1996. 25(1): p. 130-3.
11. Payne, W.E. and J.I. Garrels, Yeast protein database (YPD): a database for the complete proteome of Saccharomyces cerevisiae. *Nucleic Acids Res.*, 1997. 25(1): p. 57-62.
12. Cavin Perier, R., T. Junier, and P. Bucher, The Eukaryotic Promoter Database EPD. *Nucleic Acids Res.*, 1998. 26(1): p. 353-7.
13. Altschul, S.F., et al., Basic local alignment search tool. *J. Mol. Biol.*, 1990. 215(3): p. 403-10.
14. McEntyre, J., Linking up with Entrez. *Trends Genet*, 1998. 14(1): p. 39-40.
15. Rashbass, J., Online Mendelian Inheritance in Man. *Trends Genet*, 1995. 11(7): p. 291-2.
16. Ohkawa, H., J. Ostell, and S. Bryant, MMDB: an ASN.1 specification for macromolecular structure. *Ismb*, 1995. 3: p. 259-67.
17. Hogue, C.W., Cn3-D: a new generation of three-dimensional molecular structure viewer. *Trends Biochem. Sci.*, 1997. 22(8): p. 314-6.

European Bioinformatics Institute (EBI)

EBI is an outstation of the European Molecular Biology Laboratory (EMBL) located at Hinxton, England. Fourteen European countries and Israel support

EMBL and its outstations. EBI's main purpose is to conduct research and provide information about bioinformatics to the world's scientific community.
As of September 1994, EBI has assumed all the activities that were previously
handled at EMBL's Data Library in Heidelberg, Germany. The EBI[1] is compa-
rable to the NCBI in the United States and is the main bioinformatics server
for the European community. Its tasks and goals are similar to those of NCBI
and include:

- Bioinformatics tracking technology
- Research and development of bioinformatics software
- Training and supporting its subscribers
- Relevant bioinformatics services

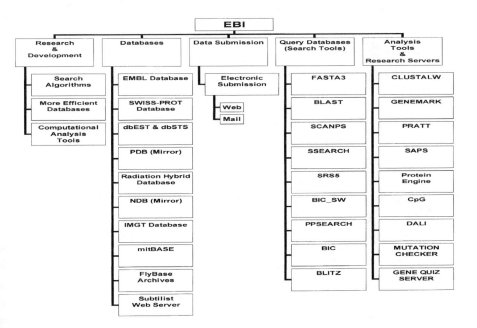

FIGURE 2.4
Overview of the European Bioinformatics Institute

Who is employed by EBI?

Like NCBI, EBI's staff consists of computer scientists, molecular biologists,
mathematicians, biochemists, research physicians, and structural biologists.
Their collaborative efforts allow them to study the molecular basis of disease
using mathematical and computational tools.

What kind of research is conducted at EBI?

The staff at EBI are involved in many facets of the bioinformatics world. Their
research tasks include:

- Developing more robust comparison algorithms
- Creating more elaborate, but user-friendly, networked information systems
- Designing more-efficient databases

What are some of the services offered by EBI?

A. Databases
B. Data submission
C. Query databases and similarity searches (e.g., FASTA[2] and BLITZ[3])
D. Online applications
E. FTP archives
F. Research and development

Databases at EBI

Following are the main databases supported by EBI's Web server:

1. EMBL database[4]
2. SWISS-PROT database[5]
3. Radiation hybrid database[6]
4. dbEST & dbSTS[7]
5. PDB[8] (Brookhaven Mirror)
6. IMGT database[9]
7. Databases on EBI ftp server
8. NDB[10] (Mirror site)
9. FlyBase archives[11]
10. MitBASE[12]
11. Subtilist Web server[13]
12. Software Bio Catalog

The first six are discussed below. Other EBI services and tools can be accessed through their website at www.ebi.ac.uk/ebi_home.html.

EMBL Nucleotide Sequence Database

This is a comprehensive database of nucleotide sequences (e.g., DNA and RNA). The nucleotide sequences at EMBL[4] are from a variety of sources. Some are from scientific literature and patent applications, but a large portion of the database includes sequences submitted directly by the sequencing source (researchers or sequencing groups). The database is a collaboration between the American Genbank nucleotide database at NCBI and the DNA database of Japan (DDBJ). The EMBL database communicates with the other two databases through its daily exchange program and constantly updates its contents. This allows EMBL to offer the worldwide scientific community an updated nucleotide database of all known public domain nucleotide sequences. In addition, EMBL's collaboration with various genomic sequencing groups allows it to introduce large-scale nucleotide sequences.

What types of information are in an EMBL nucleotide sequence file?

- The sequence
- A brief description
- Source of the sequence (the organism to which the sequence belongs)
- Bibliographic and citation information
- Locations of coding regions in the sequence (e.g., signal sequence, alpha chain, beta chain)
- Biologically significant sites in the sequence (an EST entry has very little biological information compared to a sequence entry that has been extensively studied by the researcher who reported the entry). Note: Expressed sequence tag (EST) entries are "single pass" sequences. These are typically derived from random clones and there is little functional and biological information known. It is important to know that sequences submitted by sequencing groups are extensively annotated, but the information is based on their similarity to other known sequences, not on a detailed experimental analysis of the sequence.

What are some of the genome projects that collaborate with EBI?

- Human (*H. sapiens*)
- Nematode (*C. elegans*)
- Fruitfly (*Drosophila*)
- Mouse (*M. musculus*)
- Yeast (*S. pombe*)
- Mycobacterium leprae
- Mycobacterium tuberculosis
- Methanococcus jannaschii

SWISS-PROT Protein Sequence Database[5]

The University of Geneva and EBI's EMBL Data Library together maintain the SWISS-PROT protein database. The translated DNA sequences at EMBL are directly submitted to SWISS-PROT, which is an adaptation of the PIR (Protein Identification Resource) database[14]. The SWISS-PROT database is non-redundant and has cross references to some of the other relevant libraries.

For example, its cross references to the EMBL database allow the user to gain access to the nucleotide sequence. It also has reference material from the PDB[8] and the PROSITE[15] libraries. PDB references are found only in sequence files whose 3-D structures are known and present in Brookhaven's protein data bank (PDB). The PROSITE reference is found in sequence files whose sequence contains a characterized motif present in the PROSITE Motif database.

Are all the sequence files in SWISS-PROT a result of peptide sequencing projects?
Even though the SWISS-PROT database is a protein sequence library, many of its sequence files are results of translated DNA and RNA sequence files from EMBL. The coding sequence (CDS) files at EMBL are translated into amino acid sequence files and maintained in TREMBL. TREMBL is the host of all translated EMBL nucleotide sequences and incorporates its data into the SWISS-PROT database. SP-TREMBL contains the translated sequence files, which will eventually be incorporated into the SWISS-PROT database.

How are the SWISS-PROT accession numbers assigned?
Upon submission, the EBI assigns a SWISS-PROT accession number to protein sequences whose sequence has been directly determined by a peptide-sequencing project. For translated sequence files, the accession number is assigned upon its incorporation into the TREMBL[5] library. The directly sequenced proteins can then be distinguished from those that were derived from a nucleotide sequence. A SWISS-PROT file with a cross reference to TREMBL is a certain sign of a translated nucleotide sequence. Currently, the SWISS-PROT database[5] holds about 70,000 sequence entries, while TREMBL is the host to approximately 170,000.

Radiation Hybrid Database (RHdb)
The raw data from STS, scores, experimental settings, and cross references in the RHdb database[6] are used to construct radiation hybrid maps. Currently, the database holds about one-hundred experimental conditions and approximately 75,000 radiation hybrid entries from two species (human and mouse).

What are radiation hybrid maps?
These are chromosome maps constructed from radiation hybrid score vectors, and are alternatives to genetic maps. They can be used to mark non-polymorphic markers, as well as to order unresolved clusters of polymorphic STSs. The presence of precise STS maps is invaluable for studying multifactorial genetic disorders in humans.

Can the database calculate maps or map the markers?
This site is incapable of calculating maps or mapping markers. It is a database of submitted maps that were calculated prior to their submission. The following are some of the programs used to calculate hybrid maps:

TABLE 2.1

Radiation Hybrid Database

Species	Human	Mouse
STSs	58579	316
Radiation hybrid entries	74337	328
ESTs	44623	1
CDNA sequenced (whole)	338	0
Genetic markers	4200	0
Alternative STSs (from genetic loci)	2186	
Markers in CpG islands	1	0
STSs (unknown polymorphic or expressed elements)	3331	0
Entries describing experimental conditions	107	107
Maps	23	0
Cross references to other databases	219,411	651

The reported data is as of June 6, 1998.

- RHMAP[16]
- RHMAPPER: (Whitehead Institute/MIT Center for Genome Research)
- RADMAP/MULTIMAP[17]

dbEST and dbSTS

This is a mirror database of NCBI's EST and STS entries. The database entries are composed mainly of expressed sequence tags (ESTs), single-pass cDNA sequence entries, and sequence tagged sites (STSs), short genomic landmark sequences. All of these entries are primarily maintained by NCBI.

What types of information can be retrieved from dbEST?

- Positionally cloned human disease genes
- cDNA sequence data and mapping data
- Identification of coding regions and tagging human genes

What types of information can be retrieved from dbSTS?

- More comprehensive and detailed annotations and contact data
- Experimental conditions
- Genetic map locations
- Putative homology to other entries through BLAST[18]

The EBI SRS interface can be used to search the dbEST and dbSTS libraries.

PDB (Brookhaven Mirror): Protein Data Bank

The PDB[8] library is a collection of all known, public domain 3-D structures maintained at the Brookhaven National Laboratory. This service is supported by the U.S. National Science Foundation, the U.S. Public Health Service, National Institute of Health, National Center for Research Resources, National Institute of General Medical Sciences, National Library of Medicine, and the U.S. Department of Energy. Starting in July 1999, the Protein Data Bank will no longer be maintained by Brookhaven.

What types of structures are maintained at PDB?

- Proteins
- Proteins + nucleotide sequence (e.g., DNA)
- Proteins complexed with metals
- Proteins complexed with inhibitors

How were the 3-D structures determined?

- Nuclear magnetic resonance (NMR)
- X-ray crystallography

What is the difference between the two techniques?

- In NMR, the structure of the molecule is determined in solution. This technique yields a great deal of dynamic information. We gain insight into how the molecule behaves in an aqueous (solution) environment. The structural features of the molecule give us insight into its functional characteristics.
- X-ray crystallography gives us a 3-D static picture of the molecule.The structure of the molecule is determined in its crystallized form; therefore, the structures solved through this technique lack dynamic data. In other words, we do not know how the molecule behaves in solution, its natural state.

What types of information are in a PDB file?

- Atomic coordinates determined either by NMR or X-ray crystallography
- Bibliographic citations
- Primary structure information (e.g., the amino acid sequence)
- Secondary structure information (e.g., alpha helix, beta strand)
- Crystallographic structure factors and NMR experimental data

If solving structures through NMR gives more information and insight into the molecule, why aren't all the structures solved by this method?

Solving structures through NMR is limited by the size of the molecule. Many of the proteins are outside this size range. Therefore, there is a need for alternative techniques in solving the 3-D structures of the larger molecules (e.g., X-ray crystallography).

IMGT Database: The International ImMunoGeneTics Database

This is a nucleotide database of immunologically important genes,[9] many of which belong to the immunoglobulin superfamily. Most of the molecules in the immunoglobulin superfamily are involved in immune recognition and response. T-cell receptors (TcRs), B-cell receptors (Ig), and major histocompatibility complex (MHC) molecules are some of the classic examples of the immunoglobulin superfamily.

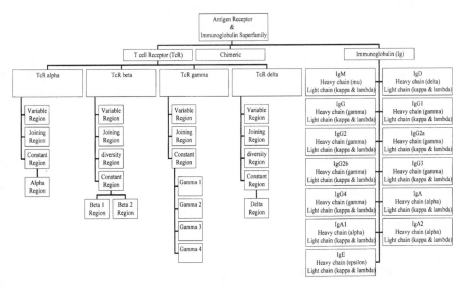

FIGURE 2.5
IMGT Classification Scheme

What types of information can be retrieved from IMGT?

- Nucleotide sequences
- Protein sequences
- Sequence alignments
- Allele, polymorphism, and STS information
- Genetic maps
- Relevance to known disorders

Are all the immunologically significant molecules at IMGT stored in one database?
IMGT data are stored in two separate databases according to their molecular identity.

1. LIGM-DB: This database maintains Immunoglobulin (Ig) and T-cell receptor (TcR) molecules. LIGM-DB stands for the Laboratoire d'ImmunoGenetique Moleculaire.[19]

2. MHC/HLA-DB: This database is primarily concerned with the major histocompatibility complex molecules.[20] In humans, these molecules are referred to as human leukocyte antigens (HLA).

Who is this service designed for and what are some of its potential applications?

- Medical researchers (e.g., HIV/AIDS research, cancer research, autoimmune disease research)

- Therapeutics and immunohistochemistry studies (e.g., antibody production and design as assay markers and potential fighting agents in certain therapies, grafts, and immunotherapy)

- Evolutionary biologists and bioinformaticians: the science of evolution and its connection to genomic diversity. The molecule's relatedness to other genes in different species could serve as a powerful tool in finding the genes of interest that are relevant in a pathogenic event.

Who are the primary contributors to the IMGT database?

- LIGM: Laboratoire d'ImmunoGenetique Moleculaire, CNRS, Université Montpellier II, Montpellier, France

- CNUSC: Centre National Universitaire Sud de Calcul, Montpellier, France

- ICRF: Imperial Cancer Research Fund, London, England

- EBI: European Bioinformatics Institute, Hinxton, England

- IFG: Institut fur Genetik, Köln, Germany

- BPRC: Biomedical Primate Research Centre, Rijswijk

- EUROGENETEC: Seraing

What are some of the tools and services offered by IMGT's server?

- Sequence alignment tools (e.g., DNAPLOT)
- Modeling tools
- Tools for mapping data
- Tools useful in classification of the query sequence
- Links to other biologically relevant databases
- Direct data submission through the Web interface

IMGT also has its own unique numbering scheme. The increased variability observed in immunologically significant molecules necessitates a more lenient or different approach in the classification and analysis of these molecules. IMGT's numbering scheme accounts for the framework (FR) of the

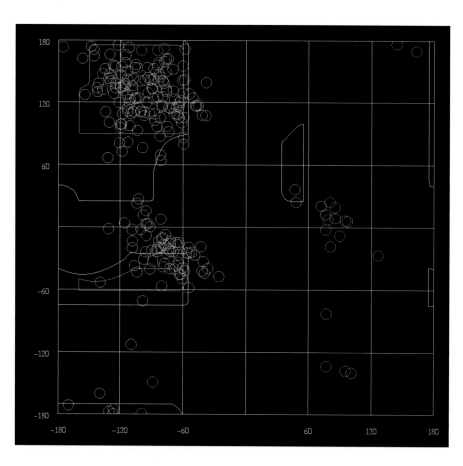

CHAPTER 1, Color Figure 1. Ramachandran Plot: Ramachandran map of 1est.pdb (accession file of Protein Data Bank). Tosyl-Elastase (E.C.3.4.21.11) is a serine protease from porcine pancreas. This enzyme hydrolyzes (cleaves) peptide bonds. The circles in the upper left corner indicate extensive beta strand formation, and the group of circles at the middle left indicate the presence of two short alpha helical secondary structures. The structure was solved in 1976 by L. Sawyer et al. (MSI).

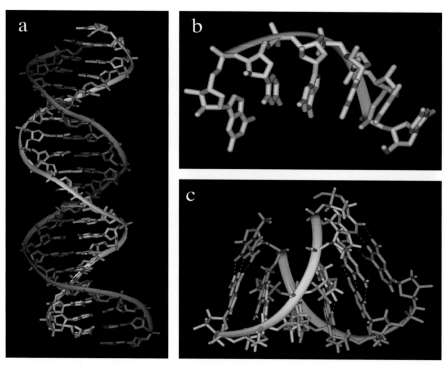

CHAPTER 1, Color Figure 2. DNA and RNA structures: a) DNA double helix--each single strand is a different color and the backbone is indicated by an idealized solid line. b) RNA single strand--free base structures point downward. c) RNA double helix with base pairs stabilized through hydrogen bond indicated by dotted lines. RNA does not form extensive double helical structures, but instead forms the more typical stem and loop motifs where a single RNA strand folds back on itself, with the stem forming a short double helix (MSI).

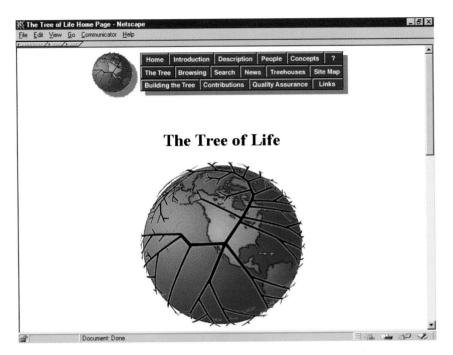

CHAPTER 3, Color Figure 1. Tree of Life Project home page at the University of Arizona, Tucson. © David R. Maddison and Wayne P. Maddison.

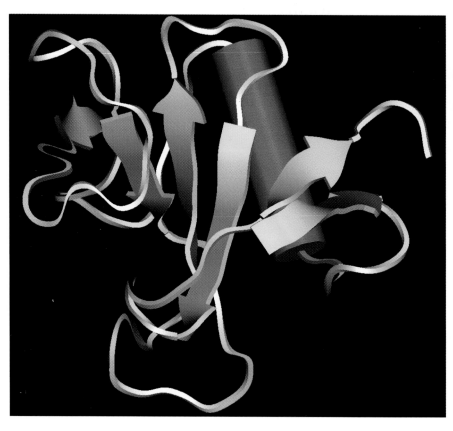

CHAPTER 5, Color Figure 1. Secondary structure illustration of penicillin acylase: a small protein with a 5-stranded anti-parallel beta sheet (yellow ribbons) interacting with a single alpha helix (red extended cylindrical structure) on its "back" side (MSI).

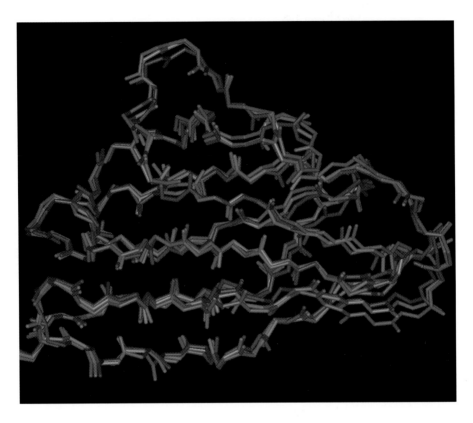

CHAPTER 5, Color Figure 2. Energy minimization of a homolog's protein structures: the homology model as it undergoes minimization. The structures are red, orange, yellow, and green. They represent the results, respectively, from the starting model: fixing splices, fixing sidechain clashes, and minimizing the whole structure (MSI).

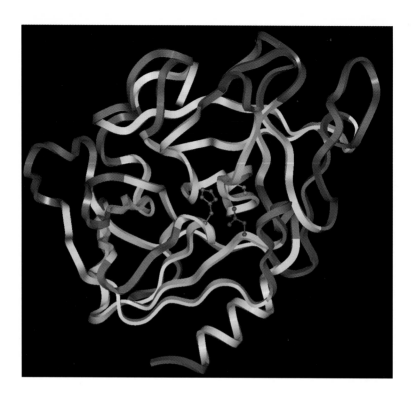

CHAPTER 5, Color Figure 3. Final model structure of serine protease: structurally conserved regions are shown in yellow, loop regions in blue, and insertions in red. The catalytic triad is also shown (MSI).

```
ACB (  E16) IVNGEEAVPGSWPWQVSLQDKT---GFHFCGGSLINENWVVTAAH (E57  )
TGD (   17) VGGYTCGANTVPYQVSLNS------GYHFCGGSLINSQWVVSAAH (57   )
EST (   16) VVGGTEAQRNSWPSQISLQYRSGSSWAHTCGGTLIRQNWVMTAAH (57   )
RP2 (  A16) IIGGVESIPHSRPYMAHLDIVTEKGLRVICGGFLISRQFVLTAAH (A57  )

ACB (  E58) CGV-TTSDVVVAGEFDQGSSSEKIQKLKIAKVFKNSKYNSL--TI (E99  )
TGD (   58) CY--KSGIQVRLGEDNINVVEGNEQFISASKSIVHPSYNSN--TL (99   )
EST (   58) CVDRELTFRVVVGEHNLNQNNGTEQYVGVQKIVVHPYWNTDDVAA (99B  )
RP2 (  A58) CKGRE--ITVILGAHDVRKRESTQQKIKVEKQIIHESYNSV--PN (A99  )

ACB ( E100) NNDITLLKLSTAASFSQTVSAVCLPSASDDFAAGTTCVTTGWGLT (E144 )
TGD (  100) NNDIMLIKLKSAASLNSRVASISLPTS--CASAGTQCLISGWGNT (144  )
EST (  100) GYDIALLRLAQSVTLNSYVQLGVLPRAGTILANNSPCYITGWGLT (144  )
RP2 (  A100) LHDIMLLKLEKKVELTPAVNVVPLPSPSDFIHPGAMCWAAGWGKT (A144 )

ACB ( E145) RY-|ANTPDRLQQASLPLLSNTNCK--KYWGTKIKDAMICAGAS- (  gap )
TGD (  145) KSSGTSYPDVLKCLKAPILSDSSCK--SAYPGQITSNMFCAGYLE (186  )
EST (  145) R-TNGQLAQTLQQAYLPTVDYAICSSSSYWGSTVKNSMVCAGG-D (186  )
RP2 (  A145) GVR-DPTSYTLREVELRIMDEKACV---DYRYYEYKFQVCVGSPT (A186 )

ACB (  gap  ) -GVSSCMGDSGGPLVCKKNGAWTLVGIVSWGSSTCS--TSTPGVY (E228 )
TGD (  187) GGKDSCQGDSGGPVVC----SGKLQGIVSWGSG--CAQKNKPGVY (228  )
EST (  187) GVRSGCQGDSGGPLHCLVNGQYAVHGVTSFVSRLGCNVTRKPTVF (228  )
RP2 (  A187) TLRAAFMGDSGGPLLC----AGVAHGIVSYGHPD----AKPPAIF (A228 )
```

CHAPTER 5, Color Figure 4. Multiple sequence alignment of four serine proteases: ACB, TGD, EST, and RP2 (MSI).

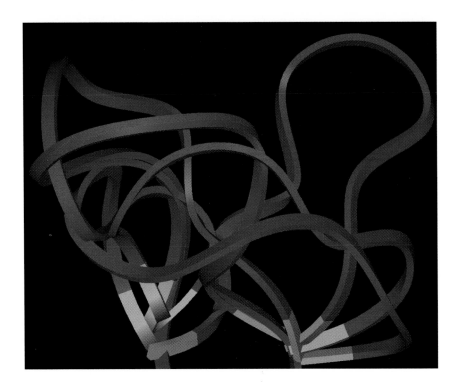

CHAPTER 5, Color Figure 5. Loop conformations located during a database search of the Protein Databank: the chosen conformation is shown in blue (MSI).

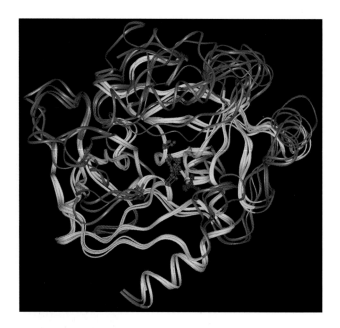

CHAPTER 5, Color Figure 6. Four serine protease structures overlaid: these are the same proteases as shown in Plate 7 (sequence alignment). The structurally conserved regions are yellow. The catalytic triad is shown in ball and stick: serine is green, histidine is orange, and aspartic acid is red (MSI).

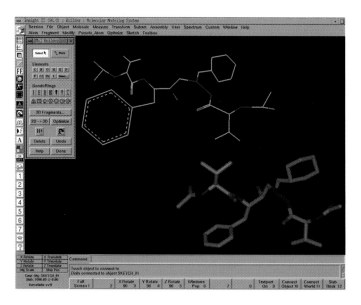

CHAPTER 5, Color Figure 7. Molecular modeling desktop: sketching a small molecule in 2D, left, that can then be automatically converted into 3D, right (MSI).

molecule, its complementarity determining regions (CDRs), its structural data when applicable, and the characterization of its hypervariable loops.

Why is this numbering scheme useful?

- It facilitates sequence comparison between the less-conserved regions of the molecules (e.g., Ig or TcR variable chains).
- The position of the conserved residues remains constant (e.g., Leu 89).
- Framework residues and residues of CDRs of same length maintain their position and can easily be identified in the absence of alignment tools.
- It allows a unique characterization of the variable regions of FRs and CDRs based on their length.
- It allows a comparative approach of the germlines (V-GENE, D-SEGMENT, J-SEGMENT, and C-GENE) in identifying mutations, polymorphisms, and somatic hypermutations. The DNAPLOT alignment tool uses this numbering scheme in its comparative steps with the different germline, functional, and ORF sequence sets.

The following keyword categories can be used to search the IMGT database for a particular sequence:

- Receptor type (e.g., chimeric, T-cell receptor, immunoglobulin)
- Receptor class (e.g., TcR alpha, TcR beta, IgM, IgG)
- Chain type (e.g., TcR alpha chain, TcR beta chain, Ig heavy chain, Ig light chain, Ig kappa chain)
- Region type (e.g., TcR constant region, Ig constant region, TcR alpha constant region, Ig delta constant region)
- Descriptive keywords (e.g., Fab, Fc, lambda 5, transgene)

Other services and tools can be accessed through EBI's website at www.ebi.ac.uk/ebi_home.html.

References

1. Emmert, D.B., et al., The European Bioinformatics Institute (EBI) databases. *Nucleic Acids Res.,* 1994. **22**(17): p. 3445-9.
2. Pearson, W.R., Using the FASTA program to search protein and DNA sequence databases. *Methods Mol. Biol.,* 1994. **25**: p. 365-89.
3. Brenner, S.E., BLAST, Blitz, BLOCKS and BEAUTY: sequence comparison on the net. *Trends Genet,* 1995. **11**(8): p. 330-1.
4. Stoesser, G., et al., The EMBL nucleotide sequence database. *Nucleic Acids Res.,* 1998. **26**(1): p. 8-15.

5. Bairoch, A. and R. Apweiler, The SWISS-PROT protein sequence data bank and its supplement TrEMBL in 1999. *Nucleic Acids Res.*, 1999. **27**(1): p. 49-54.
6. Rodriguez-Tome, P. and P. Lijnzaad, The radiation hybrid database. *Nucleic Acids Res.*, 1997. **25**(1): p. 81-4.
7. Rodriguez-Tome, P., Searching the dbEST database. *Methods Mol. Biol.*, 1997. **69**: p. 269-83.
8. Sussman, J.L., et al., Protein Data Bank (PDB): database of three-dimensional structural information of biological macromolecules. *Acta. Crystallogr. D. Biol. Crystallogr.*, 1998. **54**(1 (Pt 6)): p. 1078-84.
9. Lefranc, M.P., et al., IMGT, the International ImMunoGeneTics database. *Nucleic Acids Res.*, 1998. **26**(1): p. 297-303.
10. Berman, H.M., C. Zardecki, and J. Westbrook, The nucleic acid database: a resource for nucleic acid science. *Acta. Crystallogr. D. Biol. Crystallogr.*, 1998. **54**(1 (Pt 6)): p. 1095-104.
11. FlyBase: a Drosophila database. Flybase Consortium. *Nucleic Acids Res.*, 1998. **26**(1): p. 85-8.
12. Attimonelli, M., et al., MitBASE: a comprehensive and integrated mitochondrial DNA database. *Nucleic Acids Res.*, 1999. **27**(1): p. 128-33.
13. Moszer, I., P. Glaser, and A. Danchin, SubtiList: a relational database for the Bacillus subtilis genome. *Microbiology*, 1995. **141**(Pt 2): p. 261-8.
14. Barker, W.C., et al., Protein sequence database of the protein identification resource (PIR). *Protein Seq. Data. Anal.*, 1987. **1**(1): p. 43-98.
15. Bairoch, A., P. Bucher, and K. Hofmann, The PROSITE database, its status in 1997. *Nucleic Acids Res.*, 1997. **25**(1): p. 217-21.
16. Boehnke, M., K. Lange, and D.R. Cox, Statistical methods for multipoint radiation hybrid mapping. *Am. J. Hum. Genet.*, 1991. **49**(6): p. 1174-88.
17. Matise, T.C., M. Perlin, and A. Chakravarti, Automated construction of genetic linkage maps using an expert system (MultiMap): a human genome linkage map [published erratum appears in *Nat. Genet.* 1994 Jun;7(2):215]. *Nat. Genet.*, 1994. **6**(4): p. 384-90.
18. Altschul, S.F., et al., Basic local alignment search tool. *J. Mol. Biol.*, 1990. **215**(3): p. 403-10.
19. Lefranc, M.P., et al., LIGM-DB/IMGT: an integrated database of Ig and TcR, part of the immunogenetics database. *Ann. N.Y. Acad. Sci.*, 1995. **764**: p. 47-9.
20. Newell, W.R., J. Trowsdale, and S. Beck, MHCDB — database of the human MHC. *Immunogenetics*, 1994. **40**(2): p. 109-15.

GenomeNet — Japanese Bioinformatics Servers

GenomeNet[1] is a Japanese network of database and computational services for genome research and related research areas in molecular and cellular biology. It was established in September 1991 under the Human Genome Program (HGP) of the Ministry of Education, Science, Sports, and Culture (MESSC). GenomeNet services are operated jointly by the Supercomputer Laboratory (SCL), Institute for Chemical Research (ICR), Kyoto University and the Human Genome Center (HGC), in the Institute of Medical Science, of the University of

Tokyo. GenomeNet can be accessed at GenomeNet Services at http://www. genome.ad.jp/ and provides the following services:

DBGET/LinkDB Integrated Database Retrieval System[2]

- DBGET/LinkDB/KEGG database links diagram
- Database release information (updated daily)
- Growth of major databases (since 1982)

KEGG: Kyoto Encyclopedia of Genes and Genomes[3]

- KEGG table of contents
- Complete genomes in KEGG
- LIGAND chemical database for enzyme reactions
- BRITE: biomolecular relations in information transmission and expression
- IUPAC/IUBMB nomenclature recommendations

Sequence Interpretation Tools

- IDEAS interface to DBGET[2]/BLAST[4]/FASTA[5]

Genome Databases in Japan (Anonymous FTP of the GenomeNet)

- How to download the KEGG system

Links to Molecular Biology Servers in the World

The Japanese database system provides more than just links to DNA and pro-tein sequence databases in the U.S., Europe, and Japan. It was created to ensure the biological quality of the information content in these databases. GenomeNet was developed for the interpretation of sequence information and provides one of the most useful collections of analysis tools for a variety of biological questions. These are summarized in Figure 2.6.

GenomeNet's tools include BLAST[4] and FASTA[5], which are sequence sim-ilarity search programs, and MOTIF, a sequence motif search program devel-oped at Kyoto University. Motifs are sequence patterns rather than linear strings that correlate directly to structural features of encoded proteins. CLUSTALW[6] is a multiple sequence alignment program similar to BLAST and FASTA, which are restricted to pairwise comparison only. Other tools search for functional elements in sequences as they relate to protein structure.

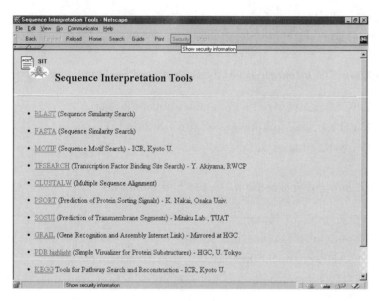

FIGURE 2.6
Sequence interpretation tools at GenomeNet (source: Kyoto Encyclopedia of Genes and Genomes, The Institute for Chemical Research, Kyoto University, www.genome.ad.jp:80/kegg/).

TFSEARCH identifies transcription factor binding sites. Transcription factors are proteins that control gene activity by directly binding to DNA close to gene sequences. Gene recognition and assembly Internet link (GRAIL) is used to identify novel genes in newly sequenced DNA for which no biological data is otherwise known.[7] Because genes have certain structures that allow their control (see transcription factors) and the expression of proteins, specific patterns or motifs of short DNA sequences (10 to 50 base pairs) are indicative of the presence of functional genes. In addition, eukaryotic genes are often split into functional and non-functional (or coding and non-coding) regions called exons and introns. Only exon DNA sequences are translated into protein sequences and thus are relevant to interpreting or predicting associated protein structures and functions. To predict these functions, programs like PSORT (prediction of protein sorting signals)[8] and SOSUI (prediction of transmembrane segments)[9] can be used. Although the majority of database information pertains to sequences, the number of protein structures is continuously increasing. This is important from an evolutionary point of view, because the structure of proteins is better conserved (it is the phenotype) than the corresponding amino acid (and DNA) sequence. The PDB highlight (simple visualizer for protein substructures) emulates the Protein Data Bank information and is used to compare structures, if available. Finally, a tool referring to the metabolic function of organisms can be found in KEGG.[3] This is a tool at Kyoto University for pathway search and reconstruction and will be explained here in more detail.

Kyoto Encyclopedia of Genes and Genomes — KEGG

KEGG is a structured database containing information about metabolic pathways in many microorganisms for which the complete genomic sequence is available, and also for those whose genome remains incomplete (e.g., human and mouse). For both the human and mouse genomes, only a fraction of their respective genome has been sequenced and many of their genes (and proteins) still remain to be characterized. The human genome, however, is projected to be finished within the next two years.

KEGG is part of the GenomeNet database system and is linked to all other publicly accessible databases by two search engines: LIGAND and BRITE.

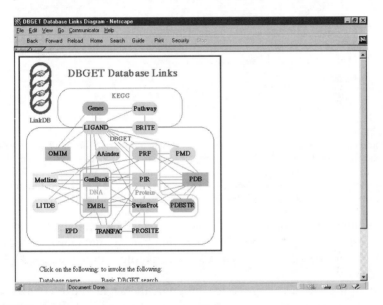

FIGURE 2.7
DBGET database links diagram: DBGET's links diagram allows easy access to the appropriate databases that contain information organized and selected for specific categories. The lines connecting the databases refer to the links between databases found within the annotated information of each entry (source: Kyoto Encyclopedia of Genes and Genomes, The Institute for Chemical Research, Kyoto University, www.genome.ad.jp:80/kegg/).

LIGAND[10] is a chemical database that allows the search for a combination of enzymes and metabolic compounds. It is maintained at the Institute for Chemical Research at Kyoto University and contains 9,317 entries: 3,705 for enzymes (enzyme reaction database), and 5,612 for metabolic compounds (chemical compound database). BRITE is a biomolecular relations information transmission and expression database at the Institute for Chemical Research of Kyoto University that contains 278 entries.

To directly search for a compound, enzyme, or gene of interest, the "Search and compute with KEGG" link on the KEGG home page will bring up a page featuring search tools for pathway maps, genome maps, coloring tools, prediction tools, and sequence similarity tools.

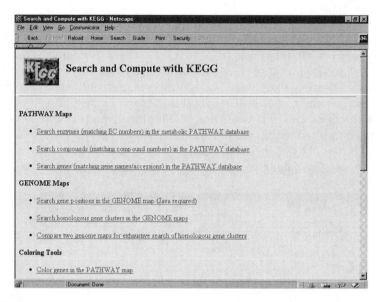

FIGURE 2.8
Search and compute with KEGG: KEGG allows the search for information about enzyme names, metabolic compounds, and genes that match the enzyme entries. KEGG provides analysis of database information based on a metabolic point of view, i.e., the information is linked in a physiologically relevant manner. (source: Kyoto Encyclopedia of Genes and Genomes, The Institute for Chemical Research, Kyoto University, www.genome.ad.jp:80/kegg/).

These links are useful only when knowing exactly what to look for because this search mode requires the exact entry number, e.g., enzyme nomenclature E.C. 2.7.1.1 for hexokinase, or chemical compound number, e.g., C00417 for cis-Aconitate.

For a more general *keyword search*, or if only the name or even partial name of an enzyme, compound, or pathway is known, the LIGAND search mode or the Pathway Maps and Molecular Catalogs link is better suited. The latter provides many search categories, among them "pathway" and "enzyme."

How to Use the KEGG Metabolic Database

How is a standard metabolic pathway found?

From the KEGG table of contents click on the "Metabolic pathways" under pathway category to see a list of all the pathways.

To find the pathway link for "lysine biosynthesis," scroll down to the group of pathways called "amino acid metabolism" and click on the link. You should now see the standard pathway MAP00300 for lysine biosynthesis.

How are species-specific pathway maps found?

Once in the standard pathway map, select the species name (e.g., *Mus musculus*) in the "Go To" window and click on Exec. You should now see the same pathway and it should show the species name in the window (e.g., *Mus musculus*).

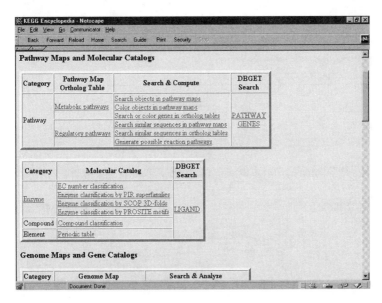

FIGURE 2.9
KEGG Encyclopedia: pathway maps and molecular catalogs (source: Kyoto Encyclopedia of Genes and Genomes, The Institute for Chemical Research, Kyoto University, www.genome.ad.jp:80/kegg/).

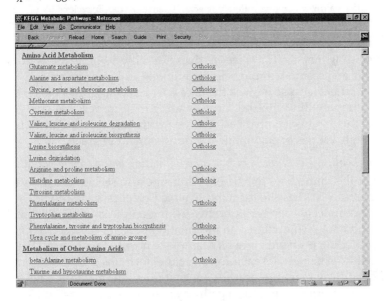

FIGURE 2.10
KEGG metabolic pathways, amino acid metabolism: the links provided allow access to the corresponding site map for the metabolisms of the amino acids listed. (source: Kyoto Encyclopedia of Genes and Genomes, The Institute for Chemical Research, Kyoto University, www.genome.ad.jp:80/kegg/).

All known mouse enzymes for which a database entry exists are labeled in green. The mouse example shows only one such enzyme (E.C. 2.3.1.-). The

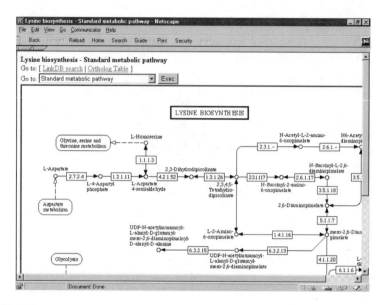

FIGURE 2.11
Lysine biosynthesis, standard metabolic pathway: the standard metabolic pathway map for
lysine biosynthesis shows all enzymes and metabolic intermediates on a generic map collected
from 33 organisms with current genome projects (completed or incomplete). This allows for
comparison of pathways among different species. (source: Kyoto Encyclopedia of Genes and
Genomes, The Institute for Chemical Research, Kyoto University, www.ge-
nome.ad.jp:80/kegg/).

corresponding map for *Homo sapiens* does not show a single marked enzyme.
L-lysine is an essential amino acid for humans and mice. L-lysine has to be
part of our diet, because we lack the necessary enzymes for its biosynthesis.[11]
Now, compare this with the corresponding map from *E. coli* and you will
find that this Gram-negative bacteria can synthesize L-lysine from L-aspar-
tate through a series of enzymatic reactions marked in green.

How is a chemical structure or metabolite information found from a pathway map?
Note that the immediate precursor of L-lysine also serves as a substrate for
peptidoglycan synthesis to form the activated precursor molecule. The structure of this molecule can be found by clicking on the small circle next to the
compound name. You will see an information page for KEGG entry C05826.
The preceding problem gives an example of how to find information about
a metabolite — namely, its chemical structure, formula, KEGG entry number,
and the pathway map numbers for which it is an intermediate. Clicking on a
substrate name (or its circle) on a pathway map page is the easiest way to find
relevant chemical information about a substrate and its shared pathways.
The same can be done for an enzyme by clicking on the E.C. number box or
for an intersecting pathway indicated in a box with smooth edges. For example, there is a link from the lysine biosynthesis map to the "lysine degradation" pathway map. Clicking on the box marked "lysine degradation" will
bring up the corresponding catabolic processes. Note that the species selec-

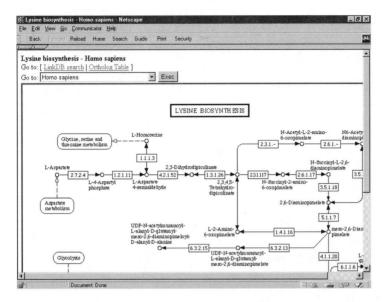

FIGURE 2.12

Lysine biosynthesis, *Homo sapiens*: same as pathway shown in Figure 2.11, except that all known enzymes in lysine metabolism are in color. No color means that the pathway does not exist or has not been described. Note that lysine is an essential amino acid and no enzymes are marked, indicating that humans do not have the capacity to synthesize this amino acid. (source: Kyoto Encyclopedia of Genes and Genomes, The Institute for Chemical Research, Kyoto University, www.genome.ad.jp:80/kegg/).

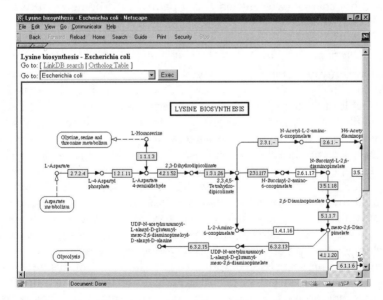

FIGURE 2.13

Lysine biosynthesis, *Escherichia coli*: lysine biosynthesis pathway for the eubacteria *E. coli*. Microorganisms can synthesize all twenty amino acids used for protein synthesis. All enzymes with database entries are in color. (source: Kyoto Encyclopedia of Genes and Genomes, The Institute for Chemical Research, Kyoto University, www.genome.ad.jp:80/kegg/).

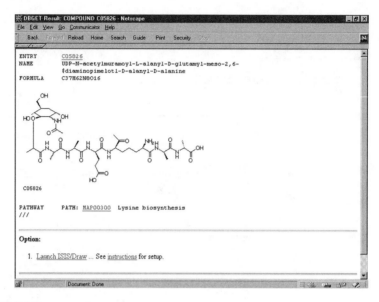

FIGURE 2.14
DBGET result, compound C05826: entry C05826 shows the chemical structure and information of UDP-N-acetylmuramoyl-L-alanyl-D-glutamyl-meso-2,6-diaminopimelotl-D-alanyl-D-alanine, a biosynthetic precursor of the bacterial cell wall component peptidoglycan. (source: Kyoto Encyclopedia of Genes and Genomes, The Institute for Chemical Research, Kyoto University, www.genome.ad.jp:80/kegg/).

tion will not change (we selected *E. coli* pathways). The new pathway map number is MAP00310.

How is a chemical structure or metabolite information found by keyword search?
To find a pathway metabolite or enzyme, the table of contents offers a direct link to the DBGET Ligand database at KEGG. This search mode can be found on the "table of contents" page under the "enzyme" category, DBGET search. Click on the link "Ligand" to access a generic search mode that allows a keyword entry. Note that having an exact enzyme number or compound number is not necessary in the DBGET database. To find information about lysine or L-lysine, type in "lysine" and hit the return (enter) key. You will receive a return list with 96 hits — the search will have returned all KEGG entries that contained the word "lysine" anywhere in the enzyme or compound name. The list contains 45 enzyme links (ec: x.x.x.xx) and 51 compound links (cpd: Cxxxxx), one being L-lysine (cpd:C00047) and all others being derivatives.
    Clicking on the cpd number will bring you to the chemical structure information sheet. This sheet lists compound entry numbers for L-lysine (note that D-lysine has a different entry), common name(s), formula, structure, all pathway maps that contain L-lysine as a metabolite (five maps for L-lysine including synthesis and degradation, biotin metabolism, alkaloid biosynthesis II, and aminoacyl-tRNA biosynthesis), and a list of all known enzymes that use L-lysine as a substrate.

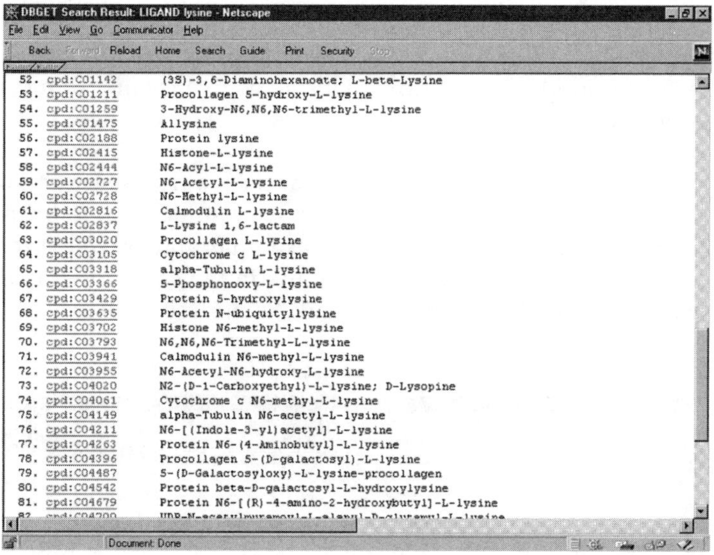

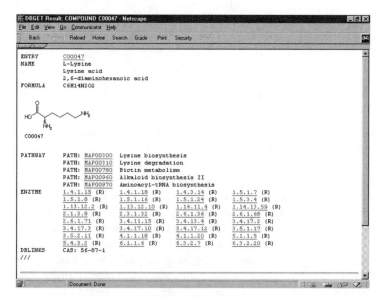

FIGURE 2.15
DBGET search result, ligand Lysine: note the long list of entries. The list contains lysines and biosynthetic derivatives thereof (exceptions are proteins in general, since all proteins contain lysine residues), as well as all enzymes involved in the metabolic pathways of these compounds. (source: Kyoto Encyclopedia of Genes and Genomes, The Institute for Chemical Research, Kyoto University, www.genome.ad.jp:80/kegg/).

FIGURE 2.16
DBGET result: Compound C00047 (Lysine) (source: Kyoto Encyclopedia of Genes and Genomes, The Institute for Chemical Research, Kyoto University, www.genome.ad.jp:80/kegg/).

64   *Bioinformatics Basics*

General Information About Biological Molecules
One additional helpful feature is the molecular catalog entry, more specifically, the "compound classification." This leads to a catalog of metabolites classified according to their functional classes, e.g., carbohydrates, fatty acids, phospholipids, neurotransmitters, etc. If you want to find the structures of a class of molecules such as amino acids or various hexoses, this link will give you the best and most comprehensive results, and can be used as a reference for structure information. If, for example, you are interested in the general structure of steroid hormones, a link in the category "Lipids" will connect to a page containing the names and chemical structures of seven cholesterol-derived steroid hormones.

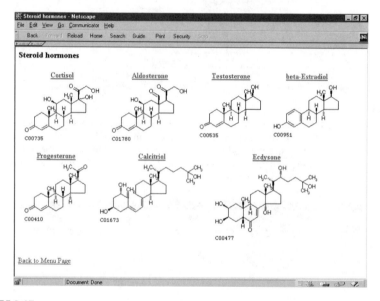

FIGURE 2.17
Steroid hormones (source: Kyoto Encyclopedia of Genes and Genomes, The Institute for Chemical Research, Kyoto University, www.genome.ad.jp:80/kegg/).

Clicking on the name link "aldosterone" connects to a structure information page providing a link to the pathway map for C21 steroid hormone metabolism (MAP00140). Following the pathway map link results in the standard pathway for steroid hormone metabolism with the aldosterone position marked as a red circle since we started our search from aldosterone. Selecting the *Homo sapiens* version of the map shows a variety of pathways, whereas the corresponding bacterial map for *E. coli* indicates — not surprisingly — that this microorganism lacks the ability to synthesize steroid hormones.

It is important to understand the limitations of databases such as KEGG. Sometimes an enzyme is not marked where you would expect it (like in the aldosterone pathway above). This pathway map shows all known reactions summarized in a standard pathway map. Species-specific enzymes are marked green. Missing enzymes that appear to interrupt a pathway occur when no entry for this enzyme (gene, amino acid sequence, protein structure) exists in any database, including KEGG. The enzyme with the entry EC 1.14.15.5 is corticosterone 18-monooxygenase and converts corticosterone into aldosterone. Following the E.C. link for this enzyme to the entry in GenBank (mirrored from NCBI) shows one nucleic acid sequence report for rat (exon 9 of rat CYP11B2 gene for aldosterone synthase). A human homolog is likely to exist for this monooxygenase, but no sequence has yet been reported.

The tools and databases described above are accessible through the Japanese GenomeNet services at http://www.genome.ad.jp.

References

1. Kanehisa, M., Linking databases and organisms: GenomeNet resources in Japan. *Trends Biochem. Sci.*, 1997. 22(11): p. 442-4.
2. Fujibuchi, W., et al., DBGET/LinkDB: an integrated database retrieval system. *Pac. Symp. Biocomput.*, 1998: p. 683-94.
3. Ogata, H., et al., KEGG: Kyoto Encyclopedia of Genes and Genomes. *Nucleic Acids Res.*, 1999. 27(1): p. 29-34.
4. Altschul, S.F., et al., Basic local alignment search tool. *J. Mol. Biol.*, 1990. 215(3): p. 403-10.
5. Pearson, W.R., Using the FASTA program to search protein and DNA sequence databases. *Methods Mol. Biol.*, 1994. 24: p. 307-31.
6. Thompson, J.D., D.G. Higgins, and T.J. Gibson, CLUSTAL W: improving the sensitivity of progressive multiple sequence alignment through sequence weighting, position-specific gap penalties and weight matrix choice. *Nucleic Acids Res.*, 1994. 22(22): p. 4673-80.
7. Roberts, L., GRAIL seeks out genes buried in DNA sequence [news]. *Science*, 1991. 254(5033): p. 805.
8. Nakai, K. and P. Horton, PSORT: a program for detecting sorting signals in proteins and predicting their subcellular localization. *Trends Biochem. Sci.*, 1999. 24(1): p. 34-6.
9. Hirokawa, T., S. Boon-Chieng, and S. Mitaku, SOSUI: classification and secondary structure prediction system for membrane proteins. *Bioinformatics*, 1998. 14(4): p. 378-9.
10. Goto, S., T. Nishioka, and M. Kanehisa, LIGAND database for enzymes, compounds and reactions. *Nucleic Acids Res.*, 1999. 27(1): p. 377-9.

2.2   Database Mining Tools

Sequence Similarity Search Tools

This chapter deals mainly with BLAST[1] and FASTA,[2] two of the most popular, user-friendly sequence similarity search tools on the Web. The BLAST server is supported through NCBI in the United States, while FASTA is maintained by the European Bioinformatics Institute (EBI) in the United Kingdom. The BLAST mirror site at EBI provides its subscribers with the option of BLAST or FASTA, along with a few other useful search programs. NCBI subscribers, however, are limited to the BLAST server which is quite effective and adaptable to various search tasks. BLAST programs are further discussed in this chapter. Additional information on alternative sequence similarity search tools can be obtained from the EBI and NCBI websites.

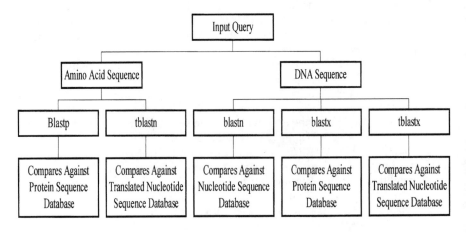

FIGURE 2.18
Overview of BLAST programs

A basic understanding of sequence alignments is necessary to comprehend BLAST or any other sequence similarity search tool. The next several pages deal with the sequence alignments on which most sequence similarity search tools are based.

What is a sequence alignment?
Sequence alignments are, for the most part, used to find potential homologs, which are then used to predict potential functions for the query sequence or to help in modeling its 3-D structure.

How are the sequence alignment tools classified?
They are classified as either global or local alignment tools.

What is a global alignment?

A global alignment[3] is the best overall alignment over the entire length of the specified sequences. The introduction of gaps within the two sequences allows for their alignment over their full length.

What are the advantages of using a global alignment tool?

The main advantage of a global alignment is its optimization of sequences that share a high degree of sequence similarity. This is useful in the alignment step of a structural modeling prediction scheme based on sequence homologs with known three-dimensional structures.

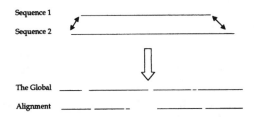

FIGURE 2.19
Schematics of a global alignment for sequences 1 and 2

What is a local alignment?

A local alignment[4] finds the optimal alignment between subregions or local regions of the specified sequences.

What are the advantages of using a local alignment tool?

A local alignment is most suitable for sequences that display localized similarity regions. A local alignment search tool is used to find sequence motifs, domains, and other types of repeats within the sequence, and is also useful for finding similar sequences for the query in a given database. In summary, a local alignment tool is best used to identify shorter regional segments with a very high degree of similarity. Often these regional segments can be used to find full-length sequence similarities.

Shared Characteristics in Both Sequence Alignment Tools

All sequence comparison algorithms rely on some sort of scoring scheme. Most use a scoring matrix to assign an overall score to each of the alignments. The alignment score is a summation of smaller scores assigned to each of its paired amino acids or nucleotide pairs. The criteria that differentiate the scoring matrices depends on the type of score it is based on. Most of the matrices are based on one of the following scoring schemes:

A scoring scheme based on "identity"

In this scoring scheme, the paired identical residues or nucleotides are assigned a positive score, while the non-identical pairs are given

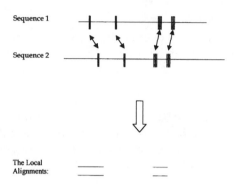

FIGURE 2.20
Schematics of local alignments for sequences 1 and 2

a score of zero. Generally, the positive score assigned to the identical pairs is one. The overall identity score is simply converted to
a percent identity.

Strengths: it is non-heuristic and simple. It works well for
sequences that have a high degree of sequence similarity.

Weaknesses: this scoring scheme is generally inferior to those that
incorporate outside knowledge. This is predominantly due to the
inequality of the non-identical pairs. For instance, an ala-
nine/valine pair is biologically more acceptable than an ala-
nine/aspartate pair. In this example, the inequality is a result of
the relative hydrophobicity of the residues involved. Therefore, the
identity-scoring scheme is less effective in detecting sequences or
sequence regions that have a low degree of sequence similarity.[5,6]
Hence, a scoring scheme that incorporates some sort of weighting
step for its non-identical pairs is biologically more significant and
effective than one that utilizes simple identity scoring. Finally, the
percent identity reported from the alignment is not always an
accurate indicator of the degree of homology that is present. This
is predominantly due to the length dependency of the percent
identity score.

A scoring scheme based on "chemical similarity"

This is basically an effort to overcome some of the limitations
associated with the identity scoring scheme. With this method, the
residue pairs are weighted according to some of their chemical and
structural characteristics. McLachlan's[7] and Feng's[5] scoring
schemes both incorporate amino acid properties such as polarity,
charge, size, and structural features into the scoring scheme.

Strengths: it agrees, to a certain extent, with the true selection pressures involved in protein structures at the amino-acid level. It is a

fact that certain mutations are more devastating to the functions of proteins than others. Generally, these mutations involve a drastic change in the characteristics of the amino acid involved. A polar to non-polar residue change or vice versa is generally more effective in altering the structure and function of the protein than a mutation involving residues with similar properties.

Weaknesses: the observed mutations in nature are not always explainable through simple scoring schemes that involve our basic understanding of such natural phenomena. Certain evolutionary mutations in nature are yet to be explained.

A scoring scheme based on the "genetic code"

This method considers the minimum number of base changes at the genomic level required to convert one amino acid into another.[8]

Strength: it is based on the principle of molecular biology.

Weakness: the element of chance produces an obstacle to the reliability of the method. A lower number of base changes does not always correspond to a greater degree of similarity between the changed residues.

A scoring scheme based on "observed mutations"

This approach is based on the frequency of mutations observed in aligned sequences.

Strength: it is based on what really happens in nature and minimizes certain otherwise faulty expectations.

Weaknesses: the scoring matrix is based on mutation frequencies found in a set of aligned sequences. The initial alignment requires human intervention which could potentially alter the true frequencies observed. Aligning the sequences by eye could introduce matching errors that would eventually lead to less-natural mutation frequencies. Scoring schemes based on observed mutations are generally better representative of natural events than those that try to explain relationships through a chemical similarity, identity, or genetic code scoring matrix.

How are sequence alignments useful?

• Phylogeny: increased sequence homology between sequences is usually an indication of a closer evolutionary realtionship.

• Structure prediction: a sequence alignment with proteins whose structures are known is used to predict the three-dimensional structure of a sequence whose structure is yet to be experimentally identified. This is based on the assumption of a direct relationship between sequence homology and structural similarity in related proteins.

- Sequence motif identification: local sequence alignments could identify potential sequence and functional motifs in proteins and nucleotide sequences.

- Function prediction: a high degree of sequence similarity between the proteins is usually an indication of shared functionality among the homologous sequences analyzed.

What is the fundamental concept behind most protein sequence algorithms?

They are based on the 210 possible pairs of amino acids that are represented by a 20x20 scoring matrix. 210 is a summation of the 20 identical and 190 mismatched amino acid pairs. The total possible pairs of characters in a given alphabet is represented through the (n-1)i formula. Hence, proteins with 20 amino acid characters have (20-1)i which corresponds to 210 possible pairs of amino acids. As discussed earlier, the identical amino acid pairs (e.g., leucine and leucine) are assigned the highest scores within the scoring matrix, followed by those that share some degree of similarity (e.g., leucine and isoleucine) and, finally, by those residues that lack similarity traits (e.g., leucine and arginine).

Basic Local Alignment Search Tool — BLAST

BLAST[1] is capable of searching all the available major sequence databases (e.g., SWISS-PROT,[9] PDB,[10] etc.). The default database in a typical BLAST run is the nr (non-redundant) database. The nr database is maintained by NCBI and its lack of redundant sequences for the same species expedites the analysis of the BLAST output file. Even though the NR database happens to be the default database for a simple BLAST run, the user still has the option of selecting the database of his choice. For instance, if the user is interested in finding protein homologs whose structures are known for modeling a protein sequence whose structure is yet to be determined, then the PDB (Protein Data Bank) database would be the most logical choice.

Who created BLAST?

BLAST's statistical theory was developed by Samuel Karlin and Steven Altschul.[1]

What type of scoring matrix is used by BLAST?

All BLAST programs use a substitution scoring matrix. The substitution matrix is used in both the scanning phase and the extension phase of the alignment process. The matrix is used to score matches. Substitution matrices are known to dramatically enhance the sensitivity of the alignment process. This is vital to BLAST's attempt to find patches of regional or local sequence similarities.

What are substitution matrices?

A substitution matrix is a scoring method used in the alignment of one residue or nucleotide against another. The first substitution matrices used in the comparison of protein sequences in evolutionary terms were developed by the late Margaret Dayhoff and her coworkers. These matrices were derived from global alignments of closely related sequences. These matrices were also used to extrapolate other matrices for less similar or evolutionarily more distant sequences.

These matrices are typically referred to as the Dayhoff, MDM, or PAM[11] set of matrices. The numbers that accompany these matrices (e.g., PAM40, PAM100, etc.) correspond to the relative evolutionary distance between the respective sequences. Smaller numbers represent evolutionarily less distant sequences, while larger numbers signify a greater evolutionary distance. The major objection to the PAM series of matrices is their incorrect assumption that the set of selection pressures on closely related sequences is the same as those that are less related. In contrast to the PAM series of matrices which are based on global alignments of closely related sequences, the BLOSUM matrices developed by Steve Henikoff and his coworkers, are derivatives of local alignments of distantly related sequences. Unlike the PAM series of matrices, the BLOSUM matrices do not extrapolate the closer-related sequences based on previously calculated matrices of a less-related set.

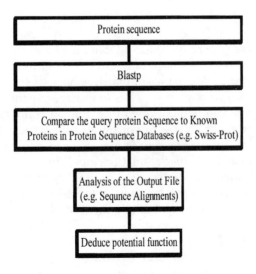

FIGURE 2.21
BLASTp diagram

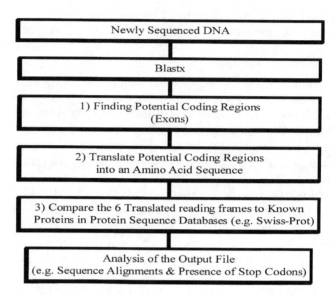

FIGURE 2.22
BLASTx diagram

The matrices in this method are all directly calculated. In contrast to the PAM series, the numbers that accompany BLOSUM matrices (e.g., BLOSUM62) refer to the minimum percent identity used to construct the matrix. Therefore, the smaller numbers correspond to blocks that are evolutionarily more distant.

Which matrix should I use?

The PAM series are generally more suitable for global similarity searches, while the BLOSUM series are found to perform better in finding regional or local similarity regions. Both series of matrices have their share of strengths and weaknesses. The current approach is to incorporate the two methods so that their strengths complement one other. Such a chimeric matrix could enhance the performance level of tomorrow's similarity searches.

BLAST programs are designed to enhance speed while maximizing the sensitivity of the distance sequence relationships. This allows the program to find the closest sequence homologs in a time-efficient manner. The BLAST programs use a heuristic algorithm that identifies local alignments. In contrast to algorithms that seek global alignments, BLAST's local alignment search finds isolated regions of sequence similarities. The BLAST server supports a variety of analysis programs that are either accessible through a web interface or can be installed on local networks to expedite the analysis step. The Standard BLAST is the original BLAST program that searches for simple sequence similarities in NCBI's database network.

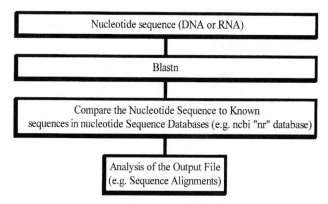

FIGURE 2.23
BLASTn diagram

What are some of the limitations associated with a basic BLAST search?

The basic BLAST program does not allow gaps in its alignments. This will, in theory, reduce the sensitivity of the search. However, the output file reports multiple regional alignments that can be used to anticipate the gaps between the query sequence and the database entry.

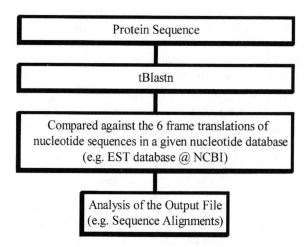

FIGURE 2.24
tBLASTn diagram

What are the different BLAST programs and how are they useful?

- BLASTp: This program allows the user to search a protein query sequence against a protein database. This can be used to find all possible sequence homologs for a given protein query sequence.

- BLASTx: This program allows the user to search a translated nucleotide sequence against a protein database. The query nucleotide sequence is initially translated in all of its six possible reading frames. This program is particularly useful in finding nucleotide sequencing errors by comparing the translated nucleotide query sequence to its potential protein homologs in a protein sequence database. The information in a BLASTx output file can also help to identify unclear nucleotides in a given reported nucleotide sequence.

- BLASTn: This program allows the user to search a nucleotide query sequence against a nucleotide database. A newly sequenced nucleotide query can be compared with itself or its homologs for identification and potential contamination of the query sequence.

- tBLASTn: This program allows the user to search translated nucleotide sequences in a given nucleotide database against a protein query sequence. The nucleotide sequences in a given nucleotide database are initially translated into each of its six possible reading frames and then are compared with the protein query sequence. This program is particularly useful in finding protein sequencing errors by comparing the protein query sequence to its potential translated nucleotide homologs in a given nucleotide database. The information in a tBLASTn output file can also help clarify unclear amino acid residues in the given query sequence. tBLASTn is similar to BLASTx in terms of its six-reading-frame translation comparison approach, but instead of a nucleotide query sequence (BLASTx) it uses a protein sequence query.

- tBLASTx: This program allows the user to search the six frame translations of a nucleotide query sequence against the six frame translations of nucleotide sequence entries in a given nucleotide database. The tBLASTx program has similarities to both BLASTx and tBLASTn programs and can be used to complement a BLASTx search.

The new BLAST programs are called BLAST 2.0. The Gapped BLAST and the PSI-BLAST[12] are just two programs supported by the new BLAST 2.0 server. The new BLAST 2.0 server has been redesigned to optimize speed and sensitivity, while adding new capabilities that allow it to support the Gapped BLAST and PSI-BLAST programs.

What is Gapped BLAST?

The Gapped BLAST algorithm allows the introduction of gaps in the alignments returned in BLAST's output file. Gaps are deletions and insertions introduced into the sequence. This prevents the similar sequence regions from being broken into segments. The heuristic input of the algorithm allows the score to reflect biological relationships associated with the alignment.

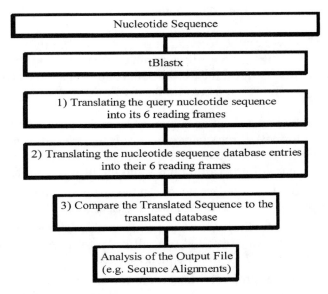

FIGURE 2.25
tBLASTx diagram

This generally reflects active sites and binding sites of the sequence that tends to be better conserved. Hence, the introduction of gaps prevents these regions from being split up into less meaningful sequence fragments.

What is PSI-BLAST?

PSI-BLAST stands for position-specific iterated BLAST.[12] The PSI-BLAST initially performs a Gapped BLAST and uses the output alignment as its input file. PSI-BLAST then constructs a position-specific score matrix that replaces the original query sequence and is used to find profiles in the next several iterative database search runs. Profile is an increased sensitivity pursuit for homologous sequences.

Some of the BLAST tools can be installed on local machines. This is the BLAST network client software. The network client software on a local computer communicates with the distant BLAST server at NCBI. BLAST2 and PowerBLAST are some of the basic network services offered through BLAST.

What is BLAST2?

BLAST2 is the standard BLAST service that provides output files in HTML format. Its filtering capabilities enable a user to find sequences with low-complexity regions.

What is PowerBLAST used for?

This is a network BLAST client that is designed to conduct large-scale analyses of genomic information. This and other network client software can be retrieved via FTP from the NCBI home page under the network directory at BLAST.

The BLAST server can be accessed in several ways, but the easiest way is through the Web (http://www.ncbi.nlm.nih.gov). A very user-friendly Web interface is used to perform the BLAST run. Following are the general steps a user needs to follow for a successful BLAST run:

1. The query sequence of interest must be in the correct format (e.g., FASTA format). If the query sequence was retrieved from NCBI's Entrez, the easiest route is to copy and paste the FASTA format of the sequence directly from Entrez into the BLAST interface.

2. The proper formatted sequence can then be pasted into the "input sequence" box on the BLAST web interface.

3. Depending on the type of sequence analyzed, the appropriate BLAST program is selected (e.g., BLASTp for protein sequences, BLASTn for DNA or RNA sequences, etc.).

4. Finally, the appropriate database must be selected. The default database on BLAST is the NCBI's nr database. The nr database will search for all the available non-redundant sequences present. For example, if the user is only interested in finding sequence homologs whose structures are known, then it would be wise to search a database that is specific to molecules with known structures. Therefore, instead of using nr, the user would select PDB[10] as the preferred database. The sequence is now ready to be submitted to the BLAST server. The results of the search can be obtained either by e-mail or seen interactively on the BLAST web interface. The e-mail route is preferable when analyzing multiple sequence files. This allows the user to analyze the sequences of interest in a time-efficient manner, while being able to analyze the result sections later (BLAST figures are saved as GIF files).

As discussed earlier, BLAST can also be accessed through the network BLAST client. In order to do this, the user must initially install the appropriate BLAST client software through FTP (ftp://ncbi.nlm.nih.gov). A BLAST search can also be done through NCBI's e-mail server (blast@ncbi.nlm.nih.gov). This is mainly for people without convenient Internet access. The query sequence must still be in the proper format in order for BLAST to conduct the proper operations. Another way to run BLAST is to install a fully executable version on the local machine that searches against the user's local databases. This version of BLAST is found under the "executables" directory of BLAST and can be obtained via FTP (ftp://ncbi.nlm.nih.gov). The executable versions of BLAST are available for IRIX6.2, Solaris2.5, DEC OSF1, and Win 32 operating systems. BLAST's output files are designed to complement other servers at NCBI. The sequences retrieved from the results section of a BLAST run are either directly or indirectly linked to NCBI's Entrez and PubMed servers.

What is the significance of the expect (E) value in a BLAST output file?

This is the number of expected matches found in a given database by chance. Therefore, the significant hits are those that are assigned lower E values. An E value of zero means that the particular hit has a zero probability of being a match by random chance. In contrast, an E value of one means that for the given database there is a probability of finding a similar scoring match by chance. The E value is basically the random noise for a given match. This value decreases exponentially with increasing score (S) values. The default E value can be increased to find statistically less-significant hits or shorter peptide or nucleotide sequences whose matching scores and E values are, in many cases, statistically less significant.

An Overview of Database Sequence Searching

Objectives:

1. Finding sequence homologs to deduce the identity of the query sequence.
2. Identifying potential sequence homologs with known three-dimensional structures for predicting the three-dimensional structure of the target sequence and deducing functional features.

Potential problems:

Being able to distinguish between a true sequence homolog (common ancestor) and one that has been found by chance in the given database can be a problem. These indistinguishable hits must undergo further examination to understand their true relationship to the query sequence.

How should a sequence database search be started?

1. A query sequence is needed. This is the target sequence that needs to be analyzed. The query sequence could be either a newly determined sequence whose identity is yet to be determined or one whose identity is known. A database search can help in determining the identity of the newly determined query sequence or finding possible sequence homologs for a known query sequence entry.
2. Select the appropriate server. The server must be reliable, regularly updated, and powerful. These characteristics are generally associated with government or government-funded bioinformatics servers such as NCBI. The National Center for Biotechnology Information (NCBI) is a collection of several public domain databases and search tools that is readily available through the Internet and is compatible with most Web browsers.
3. Select the appropriate program or set of programs in a given server. If NCBI is the server of choice and a program is needed to conduct a simple sequence similarity search, then one of the BLAST (1) programs might be appropriate.

4. Which BLAST program should be used for a simple sequence similarity search? If the query is a protein sequence, then BLASTp is the appropriate tool. If the query sequence is DNA or RNA, then the BLAST program must be utilized. These are just two of several programs available at the BLAST server. Other BLASTn programs (e.g., tBLASTn, tBLASTx) can be utilized for finding sequence homologs for the query sequence, and also to perform more advanced tasks. For instance, the BLASTx program can be used to find potential coding regions in a gene, while other BLAST programs can be utilized to check for potential sequencing errors in a newly determined sequence. (BLAST was explained in detail in the Sequence Similarity Search Tools section in Chapter 2.2).

5. Select the appropriate database. There are two ways to select the appropriate database:

   • Search all relevant databases: this is a non-redundant database of all possible submitted entries. Selecting this option will enable the user to search through all the available sequence entries.

   • Search through a specific database: in this case, the user is merely interested in a specific type of database. For instance, if the user is solely interested in finding sequence homologs with known three-dimensional structures, then the PDB (Protein DataBank) database[10] would be the most logical choice since the three dimensional structure of all its sequence entries are known.

6. Select the appropriate filter. For the convenience of its subscribers, BLAST has incorporated a set of filter options in each of its programs. The filter option excludes sequences with low complexity regions. Due to the repetitive nature of these sequences, the probability of false positive hits or random hits within the search increases and ultimately obscures the result section. It is recommended that a filter be used in a given search to reduce the number of false positives.

Can the filter option exclude a true positive from the result section?

Yes. A true positive hit with low complexity regions may be excluded from the output file. Because of this, the filter option may reduce the sensitivity of the search.

How can the sensitivity of the search be maximized while minimizing false positives?

This can be achieved by conducting two separate searches for the same query sequence. In one search, a filter is utilized to minimize false positive hits, while the other search is done without a filter to maximize sensitivity. The output files from both search results are then compared and contrasted to find possible true positive hits that were excluded from the initial filtered run.

7. Reading, comprehending, and analyzing the output file. In order to derive a possible hypothesis from the result section of the searched query, the user needs to be familiar with the terminology used in the output file. The key subjects of the output file are its assigned score for each of the found entries, and the databases and accession numbers associated with each of those entries. The score assigned to each of the sequences found is typically an indication of its homology to the query sequence. In a BLAST output file the score is also related to the expected, or E, value assigned to each entry. The E value in a BLAST output file is the probability of the sequence being a random or chance hit. The closer it is to zero, the smaller the chance of it being a random hit from a given database.

Now that the general steps in a simple sequence database search have been covered, it is important to ask the following fundamental questions:

What are the advantages of using a network database as opposed to a local database?

- Network databases are regularly updated. NCBI's collaboration with the European EBI and the Japanese DDBJ keeps their databases updated on a daily basis. These daily updates provide a dependable and non-redundant resource for their subscribers.
- Local database maintenance is not trivial and, in most cases, is beyond the scope of its average user. Utilizing and maintaining a personal database can be time consuming and expensive. These obstacles increase the value of the public domain network databases at the National Center for Biotechnology Information (NCBI), the European Bioinformatics Institute (EBI), and the DNA Database of Japan (DDBJ). (These resources were explained in detail in 2.1.)
- Network databases provide their subscribers with the appropriate search tools. The National Center for Biotechnology Information provides its subscribers with the BLAST server, as do the European Bioinformatics Institute and the DNA Database of Japan. The search tools provided by the public domain servers are regularly updated and, therefore, are more useful to their subscribers.

What are the disadvantages of using a network database instead of a local database?

- A local database is readily available in the event of a network failure.
- The subscriber is limited to the search tools provided in a network database. The scanning methods used by network servers are not always the most effective tools. Local scanning methods could be employed through a local server and could be more appropriate for that particular search.

The BLAST programs described in this chapter can be accessed through the BLAST server at www.ncbi.nlm.nih.gov (NCBI's home page).

References

1. Altschul, S.F., et al., Basic local alignment search tool. *J. Mol. Biol.*, 1990. 215(3): p. 403-10.
2. Pearson, W.R., Using the FASTA program to search protein and DNA sequence databases. *Methods Mol. Biol.*, 1994. 24: p. 307-31.
3. Huang, X., On global sequence alignment. *Comput. Appl. Biosci.*, 1994. 10(3): p. 227-35.
4. Altschul, S.F. and W. Gish, Local alignment statistics. *Methods Enzymol.*, 1996. 266: p. 460-80.
5. Feng, D.F., M.S. Johnson, and R.F. Doolittle, Aligning amino acid sequences: comparison of commonly used methods. *J. Mol. Evol.*, 1984. 21(2): p. 112-25.
6. Schwartz, R.M. and M.O. Dayhoff, Origins of prokaryotes, eukaryotes, mitochondria, and chloroplasts. *Science*, 1978. 199(4327): p. 395-403.
7. McLachlan, A.D., Repeating sequences and gene duplication in proteins. *J. Mol. Biol.*, 1972. 64(2): p. 417-37.
8. Fitch, W.M., An improved method of testing for evolutionary homology. *J. Mol. Biol.*, 1966. 16(1): p. 9-16.
9. Bairoch, A. and B. Boeckmann, The SWISS-PROT protein sequence data bank. *Nucleic Acids Res.*, 1992. 20 Suppl: p. 2019-22.
10. Sussman, J.L., et al., Protein Data Bank (PDB): database of three-dimensional structural information of biological macromolecules. *Acta. Crystallogr. D. Biol. Crystallogr.*, 1998. 54(1 (Pt 6)): p. 1078-84.
11. Wilbur, W.J., On the PAM matrix model of protein evolution. *Mol. Biol. Evol.*, 1985. 2(5): p. 434-47.
12. Altschul, S.F., et al., Gapped BLAST and PSI-BLAST: a new generation of protein database search programs. *Nucleic Acids Res.*, 1997. 25(17): p. 3389-402.

Pattern Recognition Tools and Databases

Prosite[1] is one of the most widely used databases containing biological motifs and signatures. Prosite is a collection of functional sites and sequence patterns found in many proteins.

What kind of information is stored in the Prosite database and how is it useful to its subscribers?

• Many of the characterized binding sites and motifs are gathered and maintained by Prosite. The entries in Prosite are, for the most part, cross referenced and linked to other appropriate sites. For instance, the calcium-binding EF-hand signature is well documented and its entries are further characterized by their SWISS-PROT file names. These entries are generally linked to SWISS-PROT[2] or other relevant databases.

- The Prosite file includes the sequence entries that share the matched sequence motif or signature of interest. The file will also inform the subscriber of the reported false positives, false negatives, and matched sequences whose identity is questionable.

  What is a false positive sequence? This is a sequence that contains by mere chance the signature or motif of interest. A false positive generally lacks the functional characteristics associated with the sequence motif of interest.

  What is a false negative sequence? This is a sequences that shares functionality with the true hits but lacks the specified signature sequence.

  What are the questionable sequences? These are sequences that share the motif characteristics but whose functional significance has yet to be proven through experiments. Experimentation would place these questionable sequences in either the false positive or the true positive category.

This type of information provides the investigator with a powerful tool that can enhance the efficiency of the research.

- Prosite hinders redundancy. The characterized signatures are well documented in order to minimize redundant motifs.
- Prosite has search tools for matching patterns. The PROMOT[3] search tool can be used to match a sequence against the Prosite database. It can also be used to match the sequence of interest against a set of predefined patterns. Prosearch[4] is another tool that can search the SWISS-PROT and Tremble databases with a given sequence pattern or signature. Through Prosearch, novel sequence signatures and patterns can be found efficiently in all SWISS-PROT and TREMBL sequence entries.

  How is the data presented in Prosite files? The documentation of each pattern is presented in a ".doc" file, while the actual pattern is presented in a separate file labeled ".dat". The dat file also contains information on pattern scanning programs and other sequence pattern compilations.

The Significance of Embedded Symbols Within Each Signature and How to Read and Construct Signatures

The calcium-binding EF-hand sequence motif is used as an example in order to better understand the symbols used in each of the signature sequences in the Prosite database.

The following represents Prosite's calcium-binding EF-hand signature:

D-X-[DNS]-{DENSTG}-[DNQGHRK]-{GP}-[LIVMC]-
[DENQSTAGC]-X(2)-[DE]-[LIVMFYW]

1. The hyphen is used to separate each position within the sequence motif.

2. [ ]: residues within each bracket denote the allowed residues at that particular position in the sequence motif. For instance, in [DNS] the allowed residues at that particular position are aspartate, asparagine, and serine.

3. { }: the characters within curly brackets represent residues that are not allowed for that particular position in the sequence motif. In other words, all other residues are allowed for that particular position of the sequence motif.

4. X: the letter X denotes any of the twenty amino acids.

5. (n): this denotes a repeat of a particular residue or X. For instance, X(2) can also be represented as –X-X-.

6. (n,m): this denotes a repeat of a length between n and m. For instance, A(2,5) means that a continuous stretch of two, three, four, or five alanines are all equally likely at that particular position within the sequence motif.

References

1. Bairoch, A., PROSITE: a dictionary of sites and patterns in proteins. *Nucleic Acids Res.*, 1991. 19 Suppl: p. 2241-5.
2. Bairoch, A. and B. Boeckmann, The SWISS-PROT protein sequence data bank. *Nucleic Acids Res.*, 1991. 19 Suppl: p. 2247-9.
3. Sternberg, M.J., PROMOT: a FORTRAN program to scan protein sequences against a library of known motifs. *Comput. Appl. Biosci.*, 1991. 7(2): p. 257-60.
4. Kolakowski, L.F., Jr., J.A. Leunissen, and J.E. Smith, ProSearch: fast searching of protein sequences with regular expression patterns related to protein structure and function. *Biotechniques*, 1992. 13(6): p. 919-21.

3

Genome Analysis

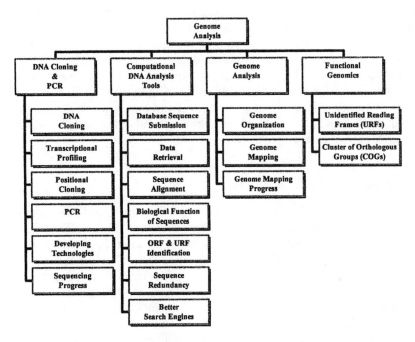

FIGURE 3.1
Chapter overview

3.1   DNA Cloning and PCR

DNA sequences are often found based on predictive tools, meaning that sequence similarities of newly discovered genes yield information about their physiological functions and structures, PCR (polymerase chain reaction), finding sequence fragments of significance (how is significance judged? By predictive biology, once again), finding distribution of genes or mRNAs (often an indicator of gene activity in an organism), and amplifying the amount of DNA for purification, sequencing, and mutational analysis.

Bioinformatics is basically database mining — the extraction, sorting, and analyzing of sequence information about genes, genomes, and proteins. Where does this information come from and how is it obtained? Simply put, genes have to be sequenced; in order to do that, they must be isolated and cloned into the appropriate form which will allow them to be manipulated in the laboratory. Therefore, cloning and sequencing are not part of bioinformatics per se. Cloning and sequencing, combined with bioinformatics, however, are interwoven activities. The technical advancement of one activity greatly impacts the other two. The increasing numbers of sequences improve the quality of statistical analysis, while the development of new bioinformatics software allows for the identification of biological functions associated with sequence patterns, thus allowing faster detection and cloning of novel genes.

Analyzing the rapidly accumulating sequence and structure information must be done with accuracy if the underlying methods of obtaining this information are to be evaluated in a critical and timely manner. Therefore, it is helpful to understand the biological background of how DNA and protein sequences are obtained.

DNA Cloning

Cloning is commonly known as the process of asexually producing a group of cells (clones), all genetically identical, from a single ancestor. Here, it refers to the use of DNA manipulation procedures to produce multiple copies of a single gene or segment of DNA through recombinant DNA technology. A desired gene or DNA fragment is cut out of its chromosomal location and inserted into *vector* DNA that is used for replication (amplification) in a host organism. Such *cloning* vectors are DNA molecules originating from viruses, bacteria, or yeast that contain proper strings of *promoter* sequences to control gene expression independently of DNA amplification in the host cell. Thus, a bacterial promoter is used on vectors for mammalian expression systems (cell lines). The bacterial RNA polymerase will specifically control the vector DNA, while the cell's genome is unaffected. Unrelated DNA fragments are integrated without loss of the vector's capacity for self replication in its natural cellular environment, allowing foreign or recombinant DNA to be reproduced in large quantities in host cells. Examples of cloning vectors are plasmids (bacterial origin), cosmids (viral origin), yeast, and bacterial artificial chromosomes (YACs, eukaryotic origin; BACs, ~150kbp inserts). Vectors are also called *expression* vectors when they contain the elements necessary for gene regulation. This feature is used to synthesize large quantities of mRNA or proteins in host organisms that normally do not contain or express these genes.

Behind every sequence stored electronically in a computer database (electronic sequencing) is a physical library of tissue samples and cloned DNA. When originating from genome projects, they are often random collections of genomic DNA obtained through shotgun techniques such as mechanical shearing or *in vitro* radiation-induced chromosome fragmentation. Chromosomal fragments are recombined into expression vectors. The collection of those cloned fragments constitutes a *library*. A vector DNA contains genes that make possible their functional expression and transformation into cell lines.

Transcriptional Profiling

How is a gene of interest selected? A gene is a sequence of base pairs, and despite the ability to predict a protein's function, experimental work to study the biochemistry and physiology of the actual protein is necessary. Yet before proceeding to laborious cell biology and biochemistry, the activity pattern of a gene (i.e., within which cells in a body and at what time of development or life stage a gene is expressed and a protein synthesized) is often the first piece of information used to select an "interesting" project or to identify a useful "target" for drug discovery. Specifically, some genes may be restricted to certain cell types, tissues, or organs; their activity patterns may change from healthy to diseased (like tumors), or may differ among young and old people.

How can the functional significance of a gene for a specific cell type, tissue, or organism be evaluated? First, cells where a gene is expressed (Northern blotting) must be found. This is done by finding the corresponding messenger RNA that serves as the intermediary template for protein synthesis (the *m* in mRNA stands for messenger, meaning that the RNA sequence is used to translate or forward the DNA sequence information into an amino acid sequence). Cytoplasmic levels (concentrations) of mRNA are good indicators for gene activity. High levels of mRNA are in many, but not all, cases indicative of the presence of a protein. The presence of protein levels must be demonstrated independently (see proteomics), if it is to be established as fact.

Identifying mRNA is done by hybridizing (binding) radioactive labeled oligonucleotides in a sequence-specific manner to isolate target mRNA. Obviously, some sequence information must be obtained in advance. This information could have been derived from short amino acid sequences obtained from protein fragments or peptides, or by searching the DNA databases for sequences with desired properties such as a human homolog to a known gene from a mouse or rat, or simply a similar, but not homologous sequence representing a potential novel gene, and so forth. Comparing the hybridization pattern of different samples at varying times during the life cycle of an organism or cell, before or after differentiation during development or under varying conditions (resting vs. hormone-induced state), can be used to construct a time–space map of where in the body a specific gene or groups of genes are actively expressed.

Once the presence of a gene of interest has been verified, DNA fragments containing the gene need to be isolated and amplified. One strategy is to use enzyme reverse transcriptase (RT) which produces a DNA copy of the mRNA fragment. The gene that codes for the mRNA can be synthesized *in vitro* and is known as *complementary* DNA, or cDNA. The cDNA represents the coding sequence of a gene including short non-coding, or flanking, regions on either side that contain regulatory sequences.

It is imperative to understand the importance of the coding sequence of eukaryotic genes because of the particularity of how most eukaryotic genes are organized on the chromosomes in the cell nucleus, which differs from the sequence found on the mRNA. A cDNA sequence of an eukaryotic gene is normally shorter than the genomic version due to the organization of genes into coding (exons) and non-coding (introns) regions. Although the entire gene (intron plus exon plus control sequences) is transcribed into an mRNA, the mRNA will be catalytically modified in order to eliminate the introns or intervening sequences. This leaves a shortened mRNA — a combination of all exon fragments found at the genome level. This is why the use of mRNA for the synthesis of cDNA yields synthetic genes that differ from their genomic origin and can be cloned into vector DNA for easier use in the laboratory (such as *in vitro* biosynthesis of proteins, transformation of DNA into new cell lines, and transgenic animals).

Positional Cloning

An alternative strategy used for the detection of hereditary disease genes is positional cloning. Here, a gene that causes a disease or contributes to the development of one, is first located on the chromosome using genetic markers, which are short, easily detectable sequences preferentially on non-coding parts of the genome. For this method, family histories must be available for analysis of the population genetics for those genes that contain mutations and appear at a specific frequency in the human population (alleles). An allele refers to a particular gene within the genome of every individual in a population. The actual sequence of the gene, however, may vary from individual to individual due to the random occurrence of mutations. While many mutations have no visible effect on the phenotype (i.e., the function of the protein), occasional mutations can cause the malfunctioning of this protein. Once a chromosomal location has been identified, clones with large inserts are identified by physical mapping, with subsequent identification of the gene(s) by sequencing. Finally, a mutation analysis compares the identified gene(s) in affected and unaffected members of a population.

Once the entire genome sequence is known, the human genome projects promise much faster identification of mutations related to diseases. Of course, the chosen sequence representing the human genome contains only one of two loci of any given gene within a population. The human genome sequence will, in fact, be the sequence of an individual revealing little information about allelic variants in a population. This will be achieved by a phenotypically selective partial genome comparison. Thus, mutation analysis of different members of the population is still a necessary step in the positional cloning approach, since no individual's genome can possibly be sequenced in its entirety. Polymorphism databases are specifically constructed to yield this information. In addition, medical databases for every disease, susceptibility for infection, cancer, and possibly psychological traits can be envisioned for the future.

NCBI provides the OMIM (Online Mendelian Inheritance in Man) database (http://www.ncbi.nlm.nih.gov/Omim/). This database is a "catalog of human genes and genetic disorders authored and edited by Dr. Victor A. McKusick and his colleagues at Johns Hopkins and elsewhere... The database contains textual information, pictures, and reference information" (NCBI). The September 1998 update list contains information about a gene related to astigmatism (OMIM entry #603047). This is a study demonstrating the difficulty of identifying familial diseases. The hereditary diseases involving ion channel proteins are discussed in Chapter 3.4. Astigmatism may serve as another example for the OMIM[1] database organization.

> In a geographically well-defined sample of 125 nuclear families of individuals affected by astigmatism, Clementi et al. (Clementi, M.; Angi, M.; Foraboso, P.; Di Gianantonio, E.; Tenconi, R.: Inheritance of astigmatism: evidence for a major autosomal dominant locus. *Am. J. Hum. Genet.* 63: 825-830, 1998.) performed complex segregation analysis by the POINTER and COMDS programs. POINTER could not distinguish between alternative genetic models, and only the hypothesis of no familial transmission could be rejected. After inclusion of the severity parameter, COMDS results defined a genetic model for corneal astigmatism and *provided evidence for single-major-locus inheritance.* These results suggested that genetic linkage studies might be feasible and that they should be limited to multiplex families with severely affected individuals. Autosomal dominant inheritance was favored. (Source: NCBI)

Before this genome analysis was performed using newly developed software, no inheritance of astigmatism could be shown and environmental factors were favored as causative agents. As late as 1989, Teikari and O'Donnell (Teikari, J. M.; O'Donnell, J. J.: Astigmatism in 72 twin pairs. *Cornea* 8: 263-266, 1989.) suggested that genetic factors did not contribute to astigmatism, leaving environmental causes as the major contributors.

Polymerase Chain Reaction (PCR)

The technique that revolutionized DNA amplification is *polymerase chain reaction*[2] (PCR) and was developed in 1985 by Kary B. Mullis who was then working at Cetus Corporation. In 1993, he received the Nobel Prize in chemistry for his contribution to molecular biology and today his technique is used in virtually every biomedical laboratory in the world. The process has been automated and machines that amplify DNA from small quantities into large ones are commercially available. The process, from template design (oligonucleotide to find a target gene sequence in the genome library) to mapping an organ for gene expression, is done by computer programs.

The increasing numbers of sequences allow the search for functional units of unknown genes. Large-scale identification of gene expression by measuring messenger RNA levels allows researchers to keep pace with the sequencing results of genome projects (public and proprietary libraries and databases). DNA sequences can be used to generate short search sequences to screen for mRNA. In an effort to increase the efficiency of finding good drug targets, the pharmaceutical industry is developing *multiarray plate* and *microchip assays*[3] where hundreds, even thousands, of gene fragments or cell types can be screened in a single assay. *DNA chip* technology developed by Affymetrix in Santa Clara, California (http://www.affymetrix.com/) is leading the technology push to determine tissue distribution of expressed genes and so-called expression sequence tags (ESTs).

The importance of PCR to bioinformatics — and the genome projects in particular — is its ability to amplify DNA without any biological information attached. This means that both coding and non-coding regions can be analyzed as long as short stretches of sequence (between 10 and 20 nucleotides long) are known. The technique is also extremely sensitive to initial sample quantity because of the enzymatically controlled DNA amplification process in non-cellular test tube solutions.

Sequencing Technologies Under Development

"A major focus of the Human Genome Project is the development of automated sequencing technology that can accurately sequence 100,000 or more bases per day at a cost of less than 50¢ per base. Specific goals include the development of sequencing and detection schemes that are faster and more sensitive, accurate, and economical. Many novel sequencing technologies are now being explored, and the most promising ones will eventually be developed for widespread use. Second-generation (interim) sequencing technologies will enable speed and accuracy to increase by an order of magnitude (i.e., ten times greater) while lowering the cost per base. Some important disease genes will be sequenced with technologies such as high-voltage capillary and ultra-thin electrophoresis to increase fragment separation rate, and the use of resonance ionization spectroscopy to detect stable isotope labels. Third-generation, gel-less sequencing technologies aiming to increase efficiency by several orders of magnitude are expected to be used for sequencing most of the human genome. These developing technologies include enhanced fluorescence detection of individually labeled bases in flow cytometry; direct reading of the base sequence on a DNA strand using scanning tunneling or atomic force microscopy; enhanced mass spectrometric analysis of DNA sequence; and sequencing by hybridization to short panels of nucleotides of known sequence. Large-scale pilot sequencing projects will provide opportunities to improve current technologies and will reveal challenges investigators may encounter in larger-scale efforts." (from: *Primer on Molecular Genetics*, Dennis Casey, Dept. of Energy, 1992, http://www.bis.med.jhmi.edu/Dan/ DOE/intro.html.).

Fluorescent labeling of DNA fragments has radically improved the speed of sequencing DNA by using the Sanger dideoxy chain termination method (Applied Biosystems, 1987; Taq cycle sequencing, 1990). This method is based on enzymes that are capable of synthesizing DNA. By adding nucleotide substrates that block the elongation process, fragments of different lengths are being produced. By carefully separating them, fragments differing in only one nucleotide in length can be distinguished, thereby allowing us to read the sequence of the cloned DNA in its entirety.

Alternatively, the Maxam-Gilbert technique is based on enzymes that cleave the DNA clone at specific bases resulting in a mix of fragments of different lengths. Again, gel electrophoresis allows the separation of these fragments at a resolution of one nucleotide difference.

Monitoring Sequencing Progress

Many web sites contain information and links to various genome projects and databases; they all focus on specific programs and are relatively complete. It is still, however, up to the interested scientist to validate the amount and timeliness of data contained in a database. A good example of monitoring the sequencing progress of a human DNA clone can be found at the Sanger web site (www.sanger.ac.uk/; see Pruitt, K.D. 1998. *Genome Res.,* 8:4-8). The site also provides links to FTP sites through a FASTA sequence format so that a status summary for a clone or a sequence (i.e., is the protein or gene of interest cloned? Is it homologous to other species?) can be obtained.

The monitoring of progress is a fascinating activity. It is really overwhelming to see how fast the numbers of sequences are added to the many databases. For an overview, "Progress Statistics" (http://www.sanger.ac.uk/Info/Statistics/) can be accessed and will give the unfinished and finished numbers of nucleotides sequenced at the Sanger Center of the British Medical Research Council. Unfinished clones provide updated information on incomplete sequences. This allows rapid access to new genes of potential interest. Such sequence information must be dealt with cautiously, as it may contain erroneous sequences and, therefore, must be considered unpublished. It is important to note that the clone information pertains only to those obtained at the Sanger Center and does not reflect the total number of sequences for any given organism. Finished clones are annotated and submitted to GenBank, EMBL, and DDBJ (DNA Database of Japan), unfinished ones are not.

The Internet information is as varied as the people who are interested in particular projects. Unlike the three main public databases — the National Center for Biotechnology Information, the European Bioinformatics Institute, and the National Institutes of Genetics in Japan (http://www.nig.ac.jp/home.html) — an individual organization's web site generally reflects only the scope of work done there. Some people focus on the individual chromosomes of specific organisms. Some are interested in mapping entire genomes or developing resources, while others focus on automation, data handling, and analysis.

Still others are involved in developing new tools to analyze sequences, compare genomes, study the structure and expression of genes, identify polymorphism, and study chromatin structure in relation to function. All these studies put together will help generate new insights into the biological function of genomes.

References

1. Schorderet, D.F., Using OMIM (On-line Mendelian Inheritance in Man) as an expert system in medical genetics. *Am. J. Med. Genet.*, 1991. 39(3): p. 278-84.
2. Mullis, K., et al., Specific enzymatic amplification of DNA in vitro: the polymerase chain reaction. *Cold Spring Harb. Symp. Quant. Biol.*, 1986. 51(Pt 1): p. 263-73.
3. Yershov, G., et al., DNA analysis and diagnostics on oligonucleotide microchips. *Proc. Natl. Acad. Sci. U.S.*, 1996. 93(10): p. 4913-8.

3.2   Computational Tools for DNA Sequence Analysis

Classical examples of the roles of computers in life sciences are sequencing, sequence analysis, comparison, evolution, tracking mutations, finding similarities for drug design, predicting the function of proteins, and predicting the roles of genes in cellular mechanisms and pathogenesis.

The usefulness of centralized databases is not only in their availability to scientists who want to learn about other researchers' cloning efforts, but also in the ability to use them as a basis for comparative genetics. Understanding the evolution of life is impossible without an understanding of the relationship between DNA sequences of different proteins and organisms. The goal of bioinformatics, in a nutshell, is the organization of sequence databases with bibliographic and biological annotations, and the support via software for the alignment of sequences, the identification of genes, the translation of DNA sequences into amino acid sequences, and the search for homologs (evolutionary related sequences). This means the collection, storage, organization, and annotation of raw data and the construction of secondary and tertiary databases.

Fifteen years ago, it was not uncommon to find people reading DNA or amino acid sequences to each other over the telephone, thereby causing an estimated "mutation rate" that far exceeded that of natural DNA replication or transcription processes. Today, downloading a file from GenBank or SWISS-PROT is extremely easy, fast, and virtually error free.

Database Submissions

The main sources of sequence information in central databases are scientists themselves. The current development of the Internet has made the process of submitting information to NCBI, EBI, or DDBJ very easy. BankIt (World Wide Web direct deposit) or Sequin (a stand-alone program) are provided by NCBI to send sequence information and biological annotation to GenBank's staff scientists who assign them accession numbers for immediate release to the public (usually within 48 hours). Daily exchanges of new submission data between GenBank, EBI, and DDBJ ensure that the information submitted by the researcher is non-redundant (submitted only once).

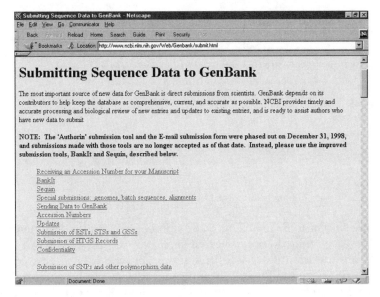

FIGURE 3.2
Submitting sequence data to GenBank

Authors are able to update their original information. Normally, scientists have a single sequence of a gene they have discovered and some relevant biological information. The genome projects, however, require specialized submission procedures for sequence information originating from ESTs (expressed sequence tags), STSs (sequence tagged sites), and GSSs (genome survey sequences). These sequences differ from traditional sequences of functional genes or proteins in their relative short length and large number.

ESTs are short sequences of 300-500bp and represent actually expressed genes because they are obtained through RT-PCR of messenger RNA extracted from tissues and cells. In addition to their proper sequence, these short sequence tags are markers that are helpful in locating (map) genes on chromosomes. EST submissions, therefore, include both sequence and mapping information. They are often submitted in batches from dozens to thousands and contain redundant information with regards to citation, submission data, and library information. GenBank provides online information about all submission requirements.

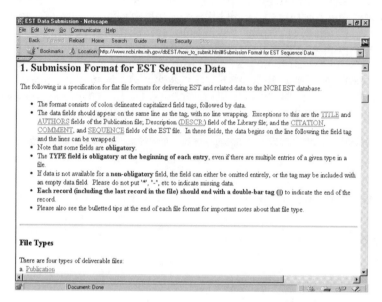

FIGURE 3.3
EST data submission: 1. Submission format for EST expressed sequence tag

STSs are similar to ESTs in length and number of submitted sequences per batch. However, they do not represent gene expression patterns but instead provide unique identifiers within a given genome identifiable by PCR. Although ESTs are the fastest growing segment of public databases, STS sequences will soon outnumber them because of the large percentage of non-coding regions of these genomes.

Because of their potential usefulness to the scientific community, genomic sequences submitted to NCBI are processed (genome center, clone name, accession number) on a daily basis, and can be submitted before they are completed. The high throughput genomic (HTG) sequences division at NCBI distinguishes three phases: 1) unfinished, unordered, 2) unfinished, ordered, and 3) high-quality finished sequences that do not contain any sequence gaps. Because of the accelerated pace of high-throughput sequencing and submission, the identification of errors is important.

To expedite this process, NCBI has set up streamlined submission procedures and deadlines to ensure the prompt and error-free release of new sequences into its ENTREZ[1] system. Without speed and accuracy, any data analysis will suffer greatly. Errors propagate very quickly through electronic media and because much of bioinformatics is used for predictive purposes (identifying novel genes, novel functions, drug discovery, predicting structures, phylogenetic relationship), errors at the sequence level will result in

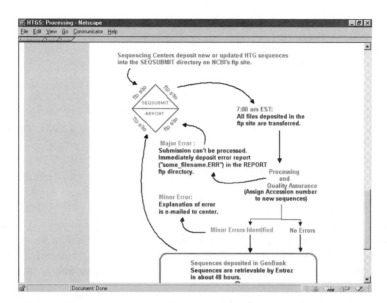

FIGURE 3.4
HTGS, processing: submission error communication procedure using FTP protocol transmission

erroneous interpretation and conclusions. The greater concern at NCBI, however, is with erroneous annotation of deposited sequences. To solve this problem, trained scientists must comb the databases and correct any mistakes. Again, the propagation of faulty annotation reduces the reliability of the data for comparative biology.

Although GenBank relies heavily on the direct submissions of individual scientists and high-throughput sequencing centers (such as Sanger Center, TIGR), NCBI's staff continuously screens biomedical journals for published sequence and structure data and uses it for annotation purposes. According to GenBank's 96.0 release notes:

> GenBank contains sequences submitted directly to the database by authors as well as entries created at NLM (National Library of Medicine; http://www.nlm.nih.gov/) derived from scanning the biomedical literature. Over 325,000 articles per year from 3400 journals are scanned for sequence data. They are supplemented by journals in plant and veterinary sciences through collaboration with the National Agricultural Library. GenBank is a component of a tri-partite, international collaboration of sequence databases in the U.S., Europe, and Japan. The collaborating databases in Europe are the European Molecular Biology Laboratory (EMBL) at Hinxton Hall, UK, and the DNA Database of Japan (DDBJ) in Mishima, Japan. Sequence data is also incorporated from the Genome Sequence DataBase (GSDB), Santa Fe, NM. Patent sequences are incorporated through arrangements with the U.S. Patent and Trademark Office, and via the collaborating international databases from other international patent offices. The database is converted to various output formats, including the Flat File and Abstract Syntax Notation 1 (ASN.1) versions.

> The ASN.1 form of the data is included on the Entrez: Sequences CD-ROM and is also available, as is the flat file, by anonymous FTP to 'ncbi.nlm.nih.gov'. (From ftp://ncbi.nlm.nih.gov/genbank/release.notes/gb96.release.notes).

To grasp the extent of today's vast computer network and the tremendous flow of genetic data, we must go back 30 years, when molecular biology was in its infancy and none of today's techniques existed. The genetic code had just been discovered and the discovery of restriction enzymes — the tools used to cut DNA — occurred soon thereafter. The sequencing of proteins was faster than sequencing nucleic acids; biochemists spent months and years establishing amino acid sequences by sequentially splitting amino acids off large amounts of purified proteins. Pioneers such as Margaret Dayhoff, one of the first biologists to make use of comparing amino acid sequences for evolutionary information, recognized the need to establish public sequence databases and her input was instrumental in developing computer-based analysis tools. As a result of Margaret Dayhoff's efforts, the first protein sequence database was established in the early 1960s.

Today, amino acid sequences are routinely obtained by means of molecular biology; i.e., by sequencing the gene first and then inferring the amino acid sequence from the DNA sequence according to the appropriate codon usage.

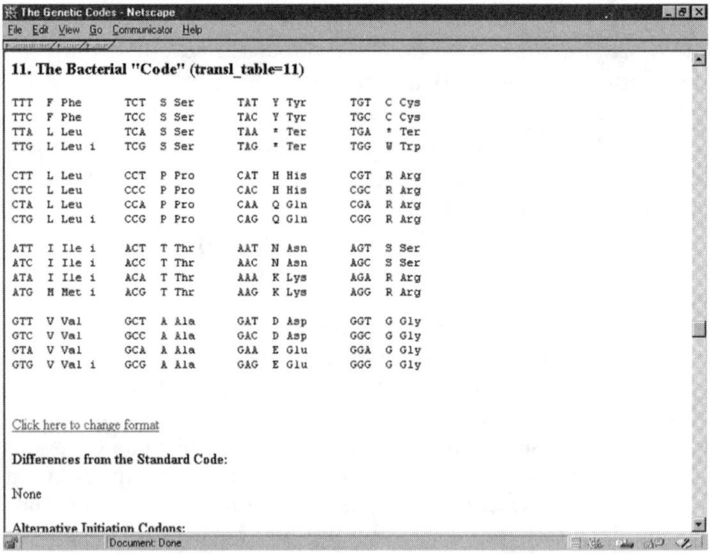

The Genetic Codes - Netscape

File Edit View Go Communicator Help

11. The Bacterial "Code" (transl_table=11)

TTT F Phe	TCT S Ser	TAT Y Tyr	TGT C Cys
TTC F Phe	TCC S Ser	TAC Y Tyr	TGC C Cys
TTA L Leu	TCA S Ser	TAA * Ter	TGA * Ter
TTG L Leu i	TCG S Ser	TAG * Ter	TGG W Trp
CTT L Leu	CCT P Pro	CAT H His	CGT R Arg
CTC L Leu	CCC P Pro	CAC H His	CGC R Arg
CTA L Leu	CCA P Pro	CAA Q Gln	CGA R Arg
CTG L Leu i	CCG P Pro	CAG Q Gln	CGG R Arg
ATT I Ile i	ACT T Thr	AAT N Asn	AGT S Ser
ATC I Ile i	ACC T Thr	AAC N Asn	AGC S Ser
ATA I Ile i	ACA T Thr	AAA K Lys	AGA R Arg
ATG M Met i	ACG T Thr	AAG K Lys	AGG R Arg
GTT V Val	GCT A Ala	GAT D Asp	GGT G Gly
GTC V Val	GCC A Ala	GAC D Asp	GGC G Gly
GTA V Val	GCA A Ala	GAA E Glu	GGA G Gly
GTG V Val i	GCG A Ala	GAG E Glu	GGG G Gly

Click here to change format

Differences from the Standard Code:

None

Alternative Initiation Codons:

Document: Done

FIGURE 3.5
The genetic code; the bacterial code: translation table 11

Amino acid sequencing, however, is still used to analyze short peptides. Its use has been boosted recently by the increased interest in proteomics. Peptide fragments obtained from protein expression profiles are micro-sequenced to determine molecular mass and charge (pI). Based on these short amino acid sequences obtained from protein extracts, protein expression profiles and posttranslational modifications can be quickly analyzed.

It was not until the 1980s that DNA sequence databases were established. The Internet is a result of the scientific community's need to transfer information quickly and reliably. Before the use of web browsers, the standard methods for downloading files from remote computers (locations) were the *file transfer protocols*, FTP and Kermit, using public domain software packages. These are still used for communication and uploading and downloading files between supercomputer centers (see Pittsburgh Supercomputing Center; http://www.psc.edu). Until 1989, the most common forms of sequence submission and retrieval were surface mail (hard copies, floppy disks, magnetic tapes), telex, and dial-up online networks. The Human Gene Mapping Library (HGML, Cold Spring Harbor Laboratories) manually updated their database annotations, and staff scientists at GenBank screened journals for published sequences. During that same period, only 50% of all entries were submitted directly by the scientists involved. Of these entries, 70% were in computer-readable form for UNIX-based Sun workstations. UNIX, of course, still has not been replaced. Internet browsers are user-friendly interface programs for Windows and Apple operating systems that transform PCs into terminals for UNIX-operated supercomputer centers and workstations.

It is sometimes more advantageous to run many stand-alone programs and downloadable data files on local stations since supercomputers are time limited, costly, and sometimes restrict access. Large companies install mirror sites on their local servers to combine public and proprietary databases for data mining. By doing this, their ongoing bioinformatics research is restricted to big business. For the individual user, interactive analysis through web browsers on remote host computers and the use of e-mail to receive results is the most common form of bioinformatics application. The increasing speed and capacity of desktop computers makes the use of stand-alone programs more feasible and reduces dependence on departmental workstations and remote servers.

1988 marked the year of the Human Genome Initiative (Alberts et al., National Academy Press) where sequences from over 1200 organisms were represented in the databases of GenBank, EMBL, and Japan. In a five-year period, the average lag time — from the publication and submission of a DNA sequence to its availability, including annotation — dropped from one year in 1983 to five months in 1988. Today, web site entry and retrieval forms provide instant access. Ten years ago, subscribers to GenBank received information on a magnetic tape every three months; CD-ROM technology was beginning to be used, but was prohibitively costly. A one-year subscription to EMBL's database cost about $200 on tape and $400 on CD-ROM for non-commercial users in the U.S. (*Methods Enzymology*, 1990, vol. 183, p. 29).

Data Retrieval

Sequence analysis includes four major biologically relevant topics: 1) comparison of gene sequences for *similarities* and defining homologies from phylogenetic analysis, 2) identification of the gene structure, including *reading frames, exon-intron* distribution, and *regulatory elements,* 3) prediction of *protein structural elements,* and 4) *genome mapping,* the linear arrangement of genes on chromosomes and its assessment within the context of metabolic pathways.

The data currently available for DNA and protein sequences is so enormous that searching for information is dubbed "biological data mining" and is analogous to mining for gold, where common rocks have to be separated from gold nuggets. Search engines perform two basic tasks: simple string searches for information retrieval of stored data (GenBank: nucleotides and proteins; and PubMed's MEDLINE: 3-D structures, genomes, and taxonomy databases), and similarity searches (e.g., BLAST) to retrieve, align, and compare sequences or structures.

The first step in sequence analysis includes retrieving sequences based on specific criteria (one of which is similarity or identity between sequences) that can be obtained through a search tool such as BLAST.[2] If no sequence is known or available, the NCBI's search engine can be screened at either the nucleotide or protein level by typing in a keyword referring to the name of a protein, the names of authors doing research on the protein of interest, or the proper accession number. These searches will retrieve reports showing the number of entries in the selected database containing the keyword anywhere in the data file.

For example, consider a researcher who is looking for bacterial proteins called porins. On the Entrez search site for "nucleotide sequence query" (http://www3.ncbi.nlm.nih.gov/Entrez/nucleotide.html), he simply types in the keyword "maltoporin" using the search field "all fields" and the search mode "automatic."

The response to the query details the existence of seven documents (Figure 3.6; March 1999) which can be retrieved by selecting "retrieve 7 documents" and selecting 20 as the number of documents to display per page (default setting). Depending on the interest in specific genes, proteins, and organisms, selecting the appropriate links will lead to the sequence information (Figure 3.7; FASTA report), annotated information (GenBank report), literature links (MEDLINE), a graphical viewer (Applet Java-based), and related protein or nucleic acid sequences. The last entry on the list with accession number M16643 refers to the *E. coli Lam*B gene encoding maltoporin, an outer membrane protein of this bacterial species that selectively facilitates transport of maltodextrins (a form of polyglucose) across this membrane. Selecting the FASTA report link shows the 5' end of the gene.

Selecting the "1 protein neighbor" link shows one entry for the maltoporin precursor protein of *Escherichia coli* (Figure 3.8, accession number AAA24059). The FASTA link shows the amino acid sequence of the N-terminal part of the

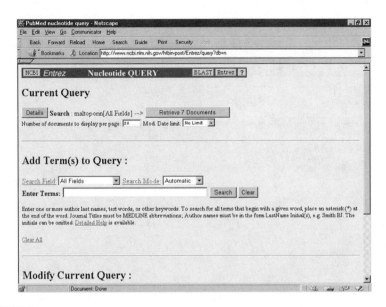

FIGURE 3.6
PubMed nucleotide query: Entrez search result for maltoporin

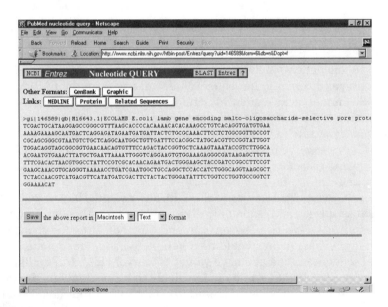

FIGURE 3.7
PubMed nucleotide query: FASTA report of maltoporin entry M16643

protein as published in the paper by Heine, Kyngdon, and Ferenci (MED-LINE link). According to the authors, the deposited partial sequence of *E. coli* maltoporin contains determinants in the *Lam*B (maltoporin) gene of *Escherichia coli* that influence the binding and pore selectivity of the protein.

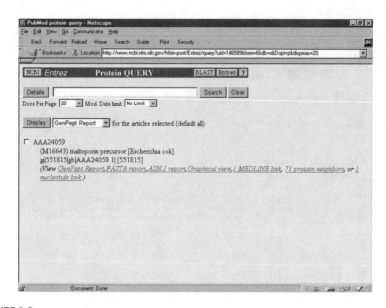

FIGURE 3.8
PubMed protein query search result for maltoporin: maltoporin Entrez accession number
AAA24059 is the equivalent of nucleotide query accession number M16643

The N-terminal amino acid sequence can now be used (FASTA format) to
BLAST (BLASTp search) the non-redundant database (GenBank CDS translations + PDB + SWISS-PROT + Spupdate + PIR, 327,237 sequences,
99,602,916 total letters) to find related sequences. This BLAST search is done
by selecting and copying the sequence information from the FASTA format
(Figure 3.9, amino acid sequence) window into the BLAST search window
and selecting *blastp* in the *Program* field, since it is a protein sequence.

The search results in 19 hits, all of which are maltoporins of *E. coli* and
related Gram-negative bacteria (*Salmonella thyphimurium*; *Yersinia enterocolit-
ica*; *Aeromonas salmonicida*; *Vibrio cholerae*; *Vibrio parahaemolyticus*; *Klebsiella
pneumoniae*).

The level of a reported similarity also indicates potential biological relationships across species and taxonomic divisions. Identities between sequences
are measured as *E-values* between zero and one indicating the chance for a
random hit. A value of one would then indicate potential randomness, while
identical sequences generally have values of zero or close to zero (like 3e-91
for entry number M16643) that are less likely to be random hits. Sequences
often show similarities below a significance threshold. A threshold E-value of
0.1 is a reasonable number for this purpose. Hits with E-values larger than
the threshold indicate a relationship that can be neither established nor
excluded *a priori*. This means that even though the DNA sequence is not suf-
ficient to establish a phylogenetic relationship between two genes, additional
information at the level of the corresponding *amino acid* sequence and *protein
structure* information (see Chapter 2.2) may nevertheless establish such a

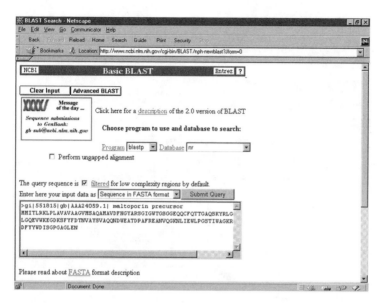

FIGURE 3.9
BLAST search for maltoporin: basic BLAST entry form using maltoporin FASTA sequence with accession number >gi551815; note that AAA24059 has the FASTA number 551815 and nucleotide entry M16643.

relationship. With an increasing number of high-resolution structures available, it can be firmly established that (super-secondary and tertiary) structures of related proteins show higher similarity than can be found by looking at the sequence alone. The reason is the redundancy in amino acid (sequence) combinations with respect to structural motifs.

Sequence Alignment

Studying a gene means understanding its variations within populations and across taxonomic groups. Sequence alignment, the pair-wise comparison of sequences, is the first step in assessing the property of a newly sequenced gene, finding homologs in other organisms, or identifying a new sequence as novel. NCBI allows us to compare two sequences using a "BLAST 2 sequences" tool. This is a specialized version of the general BLAST algorithm (see Chapter 2.2) to search for similar sequences and to retrieve them from databases. The BLAST 2 algorithm allows nucleotide (BLASTn) or amino acid sequences (BLASTp) to be compared. Several different matrix algorithms can be selected. Sequences can be entered by accession number (GI) or in FASTA format.

For multiple sequence alignment, the program ClustalW[3] is available at many bioinformatics web sites. The European Bioinformatics Institute (EBI) can be accessed for this purpose. The ClustalW interactive site can be found on the home page at http://www.ebi.ac.uk/ebi_home.html under the "Services" link and from there under "On-line applications." This site summarizes all available links provided by the EBI, as well as non-EBI servers.

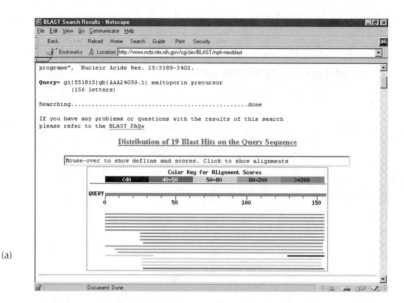

FIGURE 3.10
BLAST search result for maltoporin precursor sequence >gi551815. (a, top) Graphical JAVA-based browser interface; (b, bottom) List of sequences producing significant alignments and associated score values

Several sequences can be submitted and different output settings can be selected. The results include information about the identities from pair-wise alignments and the order of most-identical to least-identical sequence pairs. An output description for creating a graphical representation (phylogenetic tree) is also included. Programs that create trees can be downloaded as

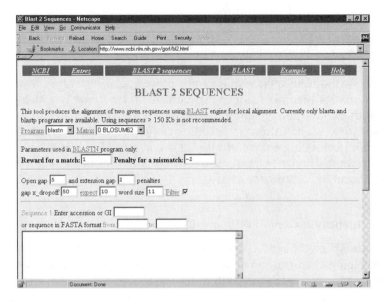

FIGURE 3.11
BLAST 2 sequences: entry form

stand-alone versions and ClustalW output files can be saved on the local hard drive for later reference, analysis, and representations.

What a Sequence Reveals About the Biological Function of a Gene

NCBI's website is an example of a matrix database for biological knowledge. To understand a gene or its sequence, its context must be known. The context for a gene means all associated biological information that defines its *function*, its structure, its cellular (chromosomal) location, the structure and function of its product — protein or RNA, and its taxonomic ranking. The following is a list of the biologically important annotations that may come with a DNA sequence:

- Related sequences in database
- Structure prediction/comparison with X-ray structure
- ORF (open reading frame) if function is unknown
- Domain structure
- Transmembrane segments
- Signal sequence
- Consensus site for glycosylation, phosphorylation, lipid anchors
- Alternative nomenclature
- Genetic information such as regulatory sequences
- Translation
- 2-D gels, pI (charge), molecular weight
- Bibliography

A recurring challenge stemming from the genome projects is identifying a DNA sequence representing or containing a gene. A gene is a functional unit in the genome of an organism. It includes regulatory sequences and a *reading frame* between a *start* and a *stop* codon, which defines the sequence corresponding to the amino acid sequence of a protein. The structure of a gene can differ dramatically from organism to organism and there are two major types: those with a continuous reading frame, and those with an interrupted reading frame (exons and introns; all exons together represent the reading frame with the introns being enzymatically cut out at the mRNA level — so-called RNA splicing). The latter are typical for higher organisms (eukaryotes) and are not found in bacteria or archaea.

How to Identify a Gene: ORFs and URFs

If a gene has been sequenced in the absence of any information about the protein, no biological function will be associated. This is an intrinsic outcome of genome projects, where long contiguous sequences of DNA have to be analyzed for the presence of genes. This requires software that identifies ORFs (open reading frames) or often URFs (unidentified reading frames) by searching for long stretches of sequence between a start and a stop codon.

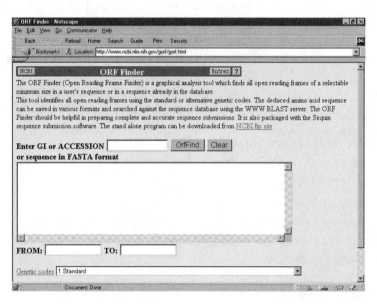

FIGURE 3.12
ORF Finder entry form: ORF finder's BLAST version also uses accession number or FASTA format of search string sequence

The length of the ORF is directly related to the size or molecular weight of the coded protein and is a useful indicator for a putative ORF. In eukaryotic genes, the signature of splice sites (i.e., the sites delineating exons and introns) provide additional help in identifying a gene. For a gene to be a functional unit, the presence of transcription consensus sequences close to a start codon must be found.

Databases provide functional sites to analyze a DNA sequence for the presence of an ORF. They allow the prediction of the associated amino acids and potential structural features of the protein. If related sequences are found, and if a related sequence already contains a gene sequence, sequence alignment — the comparison of the similarity of two or more sequences — is a good indicator for the potential biological function of the gene.

The ORF finder is a graphical analysis tool that finds all open reading frames of a selectable minimum size in a user's sequence or in a sequence that is present in the database. This tool identifies all open reading frames using the standard or alternative genetic codes. The deduced amino acid sequence can be saved in various formats and searched against the sequence database using the www.BLAST server. The ORF finder should be helpful in preparing complete and accurate sequence submissions. It is also packaged with the Sequin sequence submission software (from http://www.ncbi.nlm.nih.gov/gorf/gorf.html).

To ensure that novel genes are predicted correctly, the appropriate codon usage of the organism at hand must be carefully applied. NCBI also provides a codon usage database (http://www.ncbi.nlm.nih.gov/Taxonomy/taxonomy-home.html). The database includes the standard code for all eukaryotic organisms or taxonomic branches with the most notable exceptions given below:

- The Vertebrate Mitochondrial Code
- The Yeast Mitochondrial Code
- The Mold, Protozoan, and Coelenterate Mitochondrial Codes and the Mycoplasma/Spiroplasma Code
- The Invertebrate Mitochondrial Code
- The Ciliate, Dasycladacean, and Hexamita Nuclear Codes
- The Echinoderm Mitochondrial Code
- The Euplotid Nuclear Code
- The Bacterial "Code"
- The Alternative Yeast Nuclear Code
- The Ascidian Mitochondrial Code
- The Flatworm Mitochondrial Code
- Blepharisma Nuclear Code

The ORF finder screens a cDNA sequence for appropriate stretches flanked by a start and stop codon. "Appropriate" refers to the size of a gene, and thus the size of a protein for which no function is known or may be inferred from homolog sequences. In the latter case, the ORF finder is a tool for reliable confirmation of the identification of novel genes for known, predicted protein functions. It provides the means to ensure that the investigated cDNA sequence contains a functional version of the gene. The ORF finder is useful for screening bacterial genomes, cDNA libraries, and EST databases, but not the raw sequence of an eukaryotic organism. Therefore, the proper gene fragments, exons, and introns, must first be identified, cloned, sequenced, and put together into a contiguous sequence which then may contain the continuous coding sequence of a gene.

Genes are working units on chromosomes containing ORF fragments and non-coding regions that are important for the regulation of gene expression. Eukaryotic gene structures can often be very complex, allowing for recombinatorial processes such that a gene complex can recombine its exons in different ways (splice variants) resulting in different gene products. Entire gene clusters can form the basis of hypervariable domains in proteins of the immune system.

Software for the identification of genes from DNA fragments (sequences) is available at the Baylor College of Medicine (Gene Finder http://dot.imgen. bcm.tmc.edu:9331/gene-finder/gf.html).

A Word About Redundancy

Scientists work independently, which often results in repetitive naming of identical genes and proteins. This is typical in the submission of data for new, emerging fields; in this case, DNA sequences occur more than once and under different entries, names, and annotations. Only a specialist in the field would be able to recognize that seemingly different entries refer only to one subject. It is similar to having your name listed as three entries in a telephone book — last name, first name, and nickname — each of which refers to the same person. Obviously, redundancy can be useful; for handling DNA sequences in databases, it is actually an unintentional quality control. It sometimes happens that two competing laboratories publish the sequence of the same gene, but with one or more base differences. Does this represent a true mutant coming from different strains of mice or a sequencing error? If the source of the gene is the same organism, a sequencing error is usually the case.

Better Search Engines for Better Computers?

The way we use dictionaries and other reference books is based on our physiological pattern recognition of a set of 26 letters in alphabetical order. Of course, there could be hundreds of other alphabets that would serve the same purpose, since there are hundreds of languages. However, a language that creates words from a limited set of symbols is an optimal system for learning pattern recognition and may explain why books are such a useful, widespread, and stable form of communication and information storage. This pattern recognition may be simply referred to as analog pattern recognition as compared to digital, computer-based string searches. Computers are faster and need no alphabet (although they do use an alphabet with two symbols:

0 and 1) to find a telephone number. Hierarchical information, though essential, is not necessary for computer-based searches because search strings check the entire document. By contrast, we see information and relation in a visual representation. We know that the letter *H* comes before the letter *J*. With increasing amounts of data being stored, the inefficiency and errors (missed hits during search) increase in the absence of structured databases.

How would we teach a computer to look up a telephone number exactly the way we do, based on the alphabetical listing of last names first, first names last? We would need to teach it the rules of the alphabet as commands that the machine could interpret as a string of hierarchies and priorities when searching and categorizing information (A before B, B before C, etc.). However, we need to demand not just simple search string patterns, but a task that is much more difficult for a computer to perform: that is, for it *not* to be precise. If the name being searched for is misspelled, the computer will not find it. If you are looking up a name, however, you may be able to find it even if you are unsure about the correct spelling. In addition, you might find other information that you consider meaningful, even though you did not expect to find it. Computers and minds work very differently.

Computers are used to find information in scientific projects much the same way as they are used to find telephone numbers. Science is a human activity where finding relationships between objects is central. This means there is a need for quantification of relationships and numerical solutions for numerical relationships. Strings of information are searched for and compared, and it is the scientist who must define the quality of the string-matching process. Understanding how computers work and how scientists need the information laid out in order to extract meaning determines how successfully computers can be used in science.

→ input →

Interpretation / man   machine / computation

← output ←

Computers normally get a human interface in their everyday use in such applications as games, Internet, e-mail, word processing, and illustrations. These interfaces simulate an actual desk arrangement with stacks of paper, folders, dictionaries, and wastebaskets. The screen interface removes the necessarily abstract codes of software and machine language — a self-consistent language using symbols with precise meanings attached to them which are ultimately packaged into strings of on and off currents in electronic circuits.

Because of the widespread use of and dependence on computers in science, analog instruments have been replaced by recreating their analog interfaces on computer screens. This is a logical development since we think in analog terms, not digital. We need to see a figure, not a table of numbers. We create three-dimensional illustrations in order to understand the meaning of numerical relationships derived from scientific inquiries; for example, we use color to represent almost any physical parameter, e.g., temperature, charge density, height, stickiness, etc.

In this age of computer networks, words like "virtual cell" have new meaning. With the help of computer networks within given hierarchies of interaction, physical–chemical knowledge of cellular dynamics can be modeled mathematically. This is the world needing supercomputers for complex graphical visualization of three-dimensional objects (molecule structures) and their dynamics (molecular dynamics). The number of atoms in a single cell is literally astronomical. In computer simulations for structural movements of proteins, we not only have to include structural information such as size and relative orientation, but also the energy information of the strength of interaction between the atoms in the molecules. These are chemical bonds, dipole–dipole interactions, hydrogen bonds, and ion pairs. We are not only dealing with the number of atoms in a protein, but several physical parameters that define the interacting force between these atoms. To exemplify this complexity and the lack of computational power — as well as the lack of mathematical algorithms — we have to look at the quantum mechanical description of molecules. These systems are so complex that only the simplest molecule — molecular hydrogen — has been modeled. All other molecular structures deal with shortcuts that include mixtures of classical mechanics and quantum mechanics.

The smallest amino acid, glycine, contains ten atoms. Small proteins contain about 100 amino acids. This estimate equals several thousand atoms per protein, with several thousand proteins per cell. For the sake of simplicity, assume 10^4 atoms per protein and 10^4 proteins per cell. This corresponds to one-hundred million atoms for all proteins of a cell. The smallest genome of eubacteria is about 15 million base pairs, with an average of 70 atoms per base pair. This results in about one billion atoms per genome. Now we include all metabolites and water molecules and estimate a total of three billion atoms per single bacterial cell. If the position of every atom is defined by an average of five physical parameters, we need a spreadsheet containing 15 billion cells to store this information. We now calculate the dynamics of this system and would like to know where all the atoms are located one billionth of a second later. In order to perform this task, we need 15 billion calculation steps. Using a 100 MHz computer it would take about 150 seconds. Calculating the changes after one second would take 1.5 million days — *or more than 4,000 years.*

References

1. Schuler, G.D., et al., Entrez: molecular biology database and retrieval system. *Methods Enzymol.*, 1996. 266: p. 141-62.
2. Altschul, S.F., et al., Basic local alignment search tool. *J. Mol. Biol.*, 1990. 215(3): p. 403-10.
3. Thompson, J.D., D.G. Higgins, and T.J. Gibson, CLUSTAL W: improving the sensitivity of progressive multiple sequence alignment through sequence weighting, position-specific gap penalties and weight matrix choice. *Nucleic Acids Res.*, 1994. 22(22): p. 4673-80.

3.3 Genome Analysis

Genome analysis can determine locations of genes on chromosomes and give information on heritability and linkage to other genes, genetics (classical), medical importance, gene therapy, tracking autosomal mutations, and X-linked diseases. The yeast protein database (YPD) links information of DNA sequences, protein structure and function, cellular localization and pathways, and cell-cycle information into one coherent database with links to literature information, a commercial approach with the intention of selling database access to companies (see also Incyte Pharmaceuticals, Inc.), proteomics (Chapter 4.1) as compared to genomics, 2-D gel electrophoresis, image processing, storage, retrieval, and pattern recognition (Virage Inc. www.virage.com).

Genome Organization

Bioinformatics tools and databases are slowly becoming an integrated system that reflects the complexity of organisms. With genome projects of small organisms being completed one by one, an understanding of the differences of genomes from the three urkingdoms (eubacteria, archaea, and eukaryotes) and the relationship of genome organization to the form and function of an organism may be just around the corner. Prokaryotes have very different genome structures compared to eukaryotes. While their names refer to the absence or presence of a nuclear compartment within the cell, the differences do not stop there. The relative frequency of coding vs. non-coding regions differs as do the arrangements of genes on the chromosomes. While bacteria have a compact genome with little non-coding DNA, eukaryotic chromosomes are often extremely large and found in great numbers, especially in plants. The genes of eukaryotes and the prokaryotic archaea are often fragmented into noncontinuous "exons."

Special databases containing entire genomes of organisms provide information such as relatedness of genes within the genome, closeness in space, co-regulation, etc. For example, metabolic pathways in different bacterial species may vary because of an additional enzymatic step in one species but not another. The way to find out is to see if specific proteins belong to a cluster of genes (this structure is an *operon*) that is often aligned along the microorganism's genome such that an entire pathway for the synthesis of an amino acid is upregulated in a coordinated fashion, avoiding the individual regulation of every enzyme needed for a pathway. The existence of pathways and multiple genes coding for the enzymes of pathways has important consequences on how mutations affect cellular physiology. Mutations affecting an enzyme that is part of a pathway may affect this entire pathway because it, as such, constitutes a phenotype. The progress in sequencing entire genomes of both prokaryotic and eukaryotic organisms will undoubtedly help in determining the physiological role of organizational structures of genomes and its importance for metabolic processes.

Although genes are important because they code for all the proteins and RNA existing in a cell, these structural genes often constitute a fraction of genomes, particularly in eukaryotic organisms — fungi, plants, and animals. For example, an estimated 90% of the human genome constitutes non-coding regions. It was not that long ago that these non-coding regions were casually dismissed as junk DNA, reflecting a lack of understanding and knowledge of their function. More and more, the DNA that does not code for proteins or RNA (regulatory, structural, and enzymatic) is being recognized as important in how cells have access to the coding 10% of DNA. Essentially, this non-coding DNA is important in replication and control of cell-specific gene expression. It seems to contain information that is "read only" (short sequences that function as specific binding sites for proteins involved in gene expression and replication). Such proteins are growth factor or hormone receptors. These protein-binding elements are crucial for cells and play a role in cell differentiation, morphogenesis, and pattern formation during embryogenesis.

The implications of non-coding regions in DNA on evolution is tremendous. Because mutations are random events, the non-coding parts of chromosomes absorb most of these changes in base composition and serve as a "playground" for chromosomal recombination and accumulation of silent mutations. Polymorphic markers (the markers used in DNA fingerprinting technology) are found in this portion of the DNA. A surprising finding of genetic analysis of clusters of genes from different individuals reflects the high frequency of nucleotide sequence differences between individuals (*restriction fragment length polymorphism:* reflects sequence variations in DNA sites that can be cleaved by DNA restriction enzymes). This genetic polymorphism has recently been used in forensic science. This so-called genetic "fingerprinting" yields information unique to one individual in several billion. Genetic fingerprinting has changed our court system. The use of PCR has been successful in identification since very tiny tissue samples from blood stains, dead skin, or a single hair found at a crime scene are enough to amplify DNA for analysis.

To understand the relationship between the "blueprint" of life and life itself requires information about the relative position of genes within the genome, as well as the relationship between sequence and structure in proteins. Since proteins are not isolated entities, and multiple protein–protein interactions are the basis of cellular activity, selection pressure on individual genes is likely to be coupled over several genes whose proteins work together. This makes sequence-to-structure and structure-to-function relationships extremely complex. The multitude of interactions is too complex. New technologies such as genomics and proteomics, where the simultaneous expression levels of RNA or proteins are determined, are starting points to address the complex molecular interactions involved in cells.

How do we measure independence and interdependence of inheritable properties? We can refer to Gregor Mendel and his study of independently inherited traits on the color and consistency of peas. At the molecular level, two independent traits (phenotype) are coded for by genes (or alleles) located

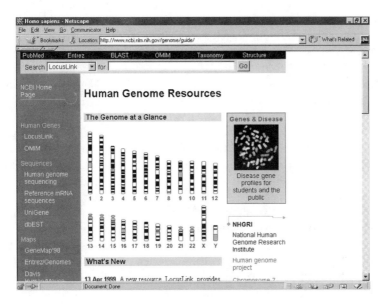

FIGURE 3.13
Homo sapiens: human genome resources, the genome at a glance

on physically separated chromosomes (Figure 3.13). If they are located on the same chromosome they are regularly — but not necessarily — inherited together (because the distance between genes on the same chromosome is also crucial); i.e., they are said not to segregate.

The importance of genome structures and chromosomal stability can be demonstrated in studies of the molecular evolution of histone proteins — the proteins responsible for the packing and storage of DNA into high-density forms of our chromosomes. During cell division, chromosomes condense into the well-known, double-arm structures (pairs of chromosomes; see karyotype), but during the normal resting state of a cell, these chromosomes are loosely packed and amenable to the proteins that transcribe genes into RNA (polymerase) and others which control access to DNA (transcription factors). This is the essence of gene regulation (transcription or expression). It is a balance of access to DNA strands among structural proteins (or histones), the nucleic acid synthesizing proteins (or polymerases), and DNA binding proteins (or transcription factors) which control accessibility of polymerases to the DNA molecule.

The importance of gene transcription in the viability of organisms is obvious, since genes code for proteins, and proteins control every process within a cell. The importance of chromosome structure, however, is less clear, but evidence indicates that interfering with the structure of chromosomes is lethal for cells. Analyzing the amino acid sequence of histone proteins has provided one line of evidence. Histone genes are highly conserved across all species within the eukaryotic kingdom. They are a key feature of the genetics of animals, plants, and fungi. Their conserved sequences also indicate a single evolutionary ancestor cell or organism from which all modern eukaryotes are derived. Indeed, histone proteins are used as molecular clocks, molecular rulers to measure the phylogenetic distance (time since separation of two species) between distantly related organisms. The survival of an organism (or a population) is linked to its phenotype, and *hereditary changes in phenotype* (mutations) occur and are stored exclusively at the DNA level by random changes in the base composition (DNA sequence). Mutations are "rejected" if the phenotype confers a lethal trait. The organism either dies before reaching reproductive age or becomes infertile, thus losing the chance to pass its genome to the next generation. The rate of mutations accumulating in a gene (nucleotide sequences which are not rejected) over time is a direct measure of the importance of the phenotype (the protein) to the viability of the individual, but not of the population. Allelic variations within a population, however, are indicators of the susceptibility of specific genes to mutations. Histone genes show an extremely low mutation rate over hundreds of millions of years indicating that the *structure* of these proteins is essential for all eukaryotic organisms. This means that besides gene replication and transcription, differential DNA packing in chromosomes during different states of the cell cycle is crucial for survival.

Genomics, the attempt to catalog the gene content, organization, and temporal expression patterns of a genome, will give detailed information about the evolution of cellular function. It is therefore not surprising that the Internet has become an essential tool for scientists because of the many databases containing information about the genomes of thousands of species, their taxonomy (Figure 3.14), and evolutionary relationship in the form of phyloge-

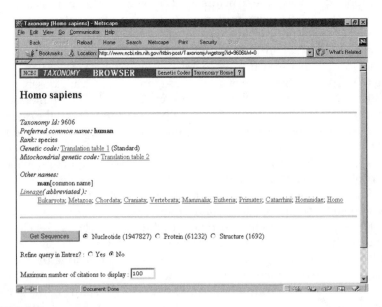

FIGURE 3.14
Taxonomy (lineage) for *Homo sapiens* (NCBI taxonomy browser)

netic trees, or "the tree of life." Phylogenetic trees are visual ways of understanding evolutionary relationships.

The tree of life is a figurative representation of the diversifying life forms on earth originating from a common ancestor. It is believed that there is only one such tree (e.g., a single progenitor "cell") and that life is not of multiple origin. Although reasonable and consistent with the findings of molecular biologists, this is speculative and corroborates the notion of the slim chance of life having arisen by chance out of non-living matter. That this event happened is not disputed here, but it is astounding since scientists agree that the spontaneous generation of new life from (in-) organic material is highly unlikely.

The Tree of Life Project at the University of Arizona (http://phylogeny.arizona.edu/tree/life.html) provides a visual overview of the phylogeny of all life on earth (see Color Figure 1*). It is not a molecular evolution type of tree, but a classical taxonomy tree. This is a very useful tool for molecular biologists who often have no formal training in evolution, zoology, botany, and ecology. The project contains information about the diversity of organisms on Earth, their history, and characteristics. It is a multi-authored website coordinated and created by David R. Maddison at the University of Arizona.

Proteins can be found as part of larger protein complexes, and only within the context of these complexes can the activity of these proteins be studied. They are not independent, thus their genes cannot be independent, yet some proteins linked through functional complex formation are coded by genes found randomly, without any apparent linkage on different chromosomes. Is there significance for such an apparent lack of organization of certain groups of genes in the human genome? The red blood cell protein hemoglobin, the transport molecule in our blood that helps carry oxygen from the lungs to target muscles or organs like the brain, is made of four tightly packed protein subunits coded for by two different genes. The genes are called alpha and beta globin genes and the functional hemoglobin protein complex contains two copies of each gene product. Although these two genes always need to be expressed together for the proper complex formation (there is no functional hemoglobin made of four alpha subunits or four beta subunits), the globin gene coding for the alpha subunit is actually a cluster of alpha globin genes with slightly different sequences and is differentially expressed during subsequent embryonic stages. Thus, only one gene copy of the alpha cluster is expressed at any given time during development. While clusters are located close to each other, the alpha subunit cluster is found on chromosome 16, while the cluster for beta globin subunits is found on chromosome 11.

Anatomical and physiological phenotypes are multi-trait phenotypes, meaning that several gene products constitute the genotype. Besides the obvious visual characteristics of an individual's appearance, cellular metabolism is the best example for studying multi-enzyme pathways. The synthesis and degradation of metabolites such as sugars, fats, amino acids, and lipids are part of these complex, interdependent pathways. The organization

---

* Color Figure 1 follows page 52.

of genes that constitute a pathway for individual metabolites in genomes is different for different organisms. As a rule, there is no strict correlation between proteins that functionally and structurally interact with each other and the position of their genes on chromosomes. Sometimes such genes are closely grouped into functional units in gene expression and are often loosely scattered all over the genome. Functional genomics may help shed some light on this problem.

There is a definite relationship between a particular DNA sequence and the chromosomal morphology of the organism. The following distinct morphological (structural) features have been established:

- Telomeric regions (tandemly repeated sequences, aging related)
- Centromeric regions (tandemly repeated sequences)
- Nucleolar organizer (genes for ribosomal RNA; relate to figure of metacentric chromosome pair)

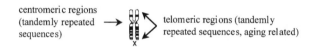

FIGURE 3.15
Chromosome pair; G-banding and chromosome identification

Because of the relationship between gene function and chromosome structure, the physical mapping of genomes is essential to gaining an understanding of the uniqueness of an organism and its developmental plan (stages of life cycle). The uniqueness of an organism lies not only in its gene composition, but also its chromosome structure. Mammalian chromosomes come in metacentric and acrocentric form. It has been shown that one reason members of different species (although closely related in their gene sequences) are reproductively incompatible is that their chromosomal structures (the superstructure of DNA which is also dependent on histone proteins) are incompatible during cell fusion and division (mitosis, meiosis). Here we can see the emergence of a loop where genes coding for histone proteins are regulated by how these proteins interact with each other and DNA to form the superstructure of chromosomes and which determines the viability of the cell because of its importance during cell division. Nucleotide changes (mutations) in histone genes affect their amino acid composition, which affects chromosome structure, which affects histone gene inheritance (replication) and expression (transcription).

Mapping the Genome

Genome databases play an increasingly important role in understanding novel genes whose functions have yet to be determined. Yet by analogy to a gene's location and association to chromosomal location, its function might be inferred and be useful in the design of future experiments. Chromosome locations, like DNA sequences, are subject to changes (e.g., mutations) and can change from generation to generation. In eukaryotes, rearrangements of chromosomal fragments (homologous recombination, reciprocal crossing over, meiosis, and mitosis) is an important part of genetic variability among individuals. Genetic polymorphism, as mentioned above, based on chromosomal rearrangements, also makes individuals genetically unique, although the overall genomic contents (the whole of all genes inherited) remain constant. Rearrangements can influence and alter gene expression in an orderly and programmed fashion.

FIGURE 3.16
GeneMap '98: a new gene map of the human genome

Many of these rearrangement processes also cause diseases, an additional motivation to understand the relationship between gene expression and chromosomal morphology. This is reflected in the growing size of genome databases related to medical issues, as well as information sites about inheritable diseases and their relationship to genetics (NIH health information at http://www.nih.gov/health/).

Genetic Linkage Maps

Genetic linkage maps depict the relative chromosomal locations of DNA markers (genes and other identifiable DNA sequences) by their patterns of inheritance. Are they inherited together or not? The distance between markers on the map indicates the frequency of how often they are inherited together (inverse relationship). This is the field of population genetics and in humans is the study of family history of autosomal and sex-linked traits. DNA markers must be polymorphic to be useful. Polymorphisms (mutations) are variations in DNA sequence that occur, on average, once every 300 to 500 bp, representing the lower end of length distribution of genes. This means that polymorphism is a fairly common feature of our genes. Mutations, however, do not necessarily translate into an altered phenotype, although many of them are responsible for observable changes such as differences in eye color, blood type, and disease susceptibility if they occur within exon sequences. It is precisely the occurrence of mutations in non-coding regions of the genome that serve as molecular markers at the level of the DNA without leaving visible traces (phenotypes) or rendering the organism less viable. Because they commonly reside in non-coding parts of the genome they can be considered *hidden mutations* and can only be "seen" at the level of DNA analysis. In short, genetic linkage maps are constructed by observing how frequently two markers are inherited together within a family tree (from generation to generation). Mendel's pea colors constitute such markers and although some are obviously inherited independently (not linked on chromosomes) while others are linked, they reside on the same chromosome.

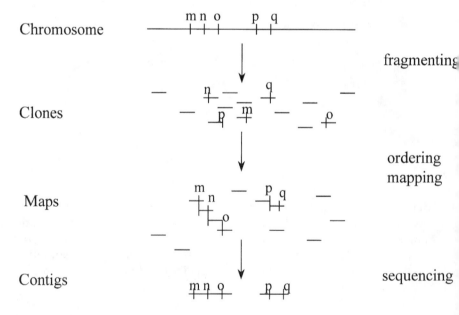

FIGURE 3.17
Mapping and sequencing of chromosomes: a chromosome containing five markers (m,n,o,p,q) is fragmentized and the fragments (clones) identified by markers. These clones are ordered and mapped by comparing overlapping strings of sequences. As long as fragments show such overlapping sequences, they can be linked to continuous segments (contig 1 mno and contig 2 pq). If some fragments are missing, the entire chromosome cannot be reproduced, but is rather represented by several contigs. Since fragments have been sequenced in full, the corresponding contig sequences are known and can be stored in the database (e.g., GenBank).

Genetic maps have been used to find the exact chromosomal location of several important disease genes, including cystic fibrosis, sickle cell disease, Tay-Sachs disease, fragile X syndrome, and myotonic dystrophy.

"One short-term goal of the genome project is to develop a high-resolution genetic map (2 to 5_cM [centimorgan]; Two markers are said to be 1_cM apart if they are separated by recombination 1% of the time. A genetic distance of 1_cM is roughly equal to a physical distance of 1 million bp (1 Mb)); recent consensus maps of some chromosomes have averaged 7 to 10_cM between genetic markers. Genetic mapping resolution has been increased through the application of recombinant DNA technology, including *in vitro* radiation-induced chromosome fragmentation and cell fusions (joining human cells with those of other species to form hybrid cells) to create panels of cells with specific and varied human chromosomal components. Assessing the frequency of marker sites remaining together after radiation-induced DNA fragmentation can establish the order and distance between the markers. Because only a single copy of a chromosome is required for analysis, even nonpolymorphic markers are useful in radiation hybrid mapping." [In meiotic mapping, described above, two copies of a chromosome must be distinguished from each other by polymorphic markers.] (from: *Primer on Molecular Genetics*, Dennis Casey, Dept. of Energy, 1992, http://www.bis.med.jhmi.edu/Dan/DOE/ intro.html).

Physical Maps

Physical maps describe the molecular organization of genes or markers within a genome or chromosome. Depending on the technique available or used, the resolution of the map can vary widely. Early methods relied on microscopic techniques analyzing banding patterns on condensed forms of chromosomes. Bands often correlate to differently active regions of genomes. While light microscopy needs DNA preparation in a fairly organized form (such as during mitosis), electron microscopy provides higher resolution and can thus detail finer structures.

High-resolution physical maps (Figure 3.18) make use of the increasing sequence information available, combining microscopic data with genetic linkage maps and DNA sequences around those markers (genes). The ultimate physical map, then, will be the entire, contiguous DNA sequence of the [human] genome or its chromosome. Since genetic linkage maps measure distances of markers based on recombinatorial activity of chromosomes, the relative distances between marker on physical and genetic linkage maps can be quite different. This is based on the fact that recombination during meiosis and mitosis has different frequency at different locations within chromosomes. The mechanism of this behavior is not understood. It could be "simply" sequence dependent or related to chromosome structure, which actually may be determined by sequence patterns. This difference between physical and functional maps, therefore, is also of interest and genome projects will contribute information that could ultimately answer their questions.

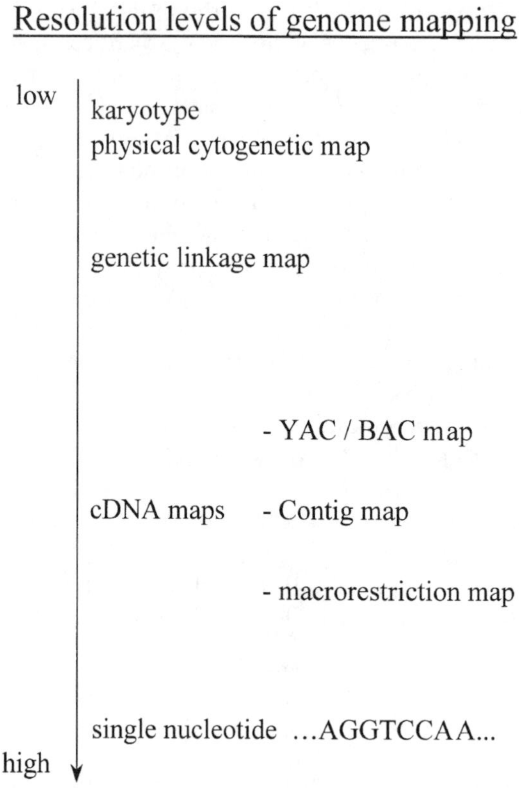

Resolution levels of genome mapping

low | karyotype
physical cytogenetic map

genetic linkage map

                              - YAC / BAC map

cDNA maps   - Contig map

                              - macrorestriction map

                    single nucleotide ...AGGTCCAA...
high ↓

FIGURE 3.18

Resolution levels of genome mapping: the genome projects progress through several levels of increasing resolution of the genetic information contained on chromosomes. The division of the genome in physically individual pieces (chromosomes) is the karyotype. Genetic markers are linked if they are on the same chromosomes (genetic linkage map). The relative position of genes can be determined by correlating physical and genetic linkage maps. Smaller fragments allow the fine resolution of markers located very close to each other on the genome in cDNA maps. Smaller and smaller fragments are amenable to complete sequence analysis at the single nucleotide level.

The Sanger web site (http://www.sanger.ac.uk/) shows examples of human chromosomes and the hierarchical organization where the user interested in a clone and its localization on a particular chromosome can zoom into an incrementally detailed map until the nucleotide sequence level is reached.

Expression Maps

The identification of structural genes has been driving the Human Genome Project from the beginning and is also characteristic of the major cloning efforts in drug discovery. The reason for this is simple: structural genes are easy to identify because they can be activated and inactivated. In fact, the problem is reduced to the identification of mRNA in cells. The resulting *sequence tags* have been instrumental in identifying novel eukaryotic genes. Instead of waiting for complete genome sequences, very short fragments are being selected and sequenced for the construction of so-called expression maps. Because genes are composed of both coding and non-coding flanking regions containing regulatory sequences, both expressed sequence tags and sequence-tagged sites have been instrumental in creating linkage and high resolution maps of human chromosomes. In the words of NCBI:

> One of the specific goals of the US Human Genome Project is the con-
> struction of a high resolution, STS (sequence tagged sites, PCR derived
> sequences from the genome) map of the genome. Up until recently, the
> great majority of STSs have been derived from anonymous genomic
> sequences. This has been hugely successful, efficient and elegant in its
> simplicity, for both genome-wide and chromosome-specific mapping.
> However, as James Sikela and coworkers proposed in 1991, the develop-
> ment of STSs from the 3′ untranslated regions (3′UTR) of mRNAs has the
> advantage of supplying not only a marker, but also a gene for the map.
> The two main advantages of using 3′UTRs rather than other portions of
> an mRNA sequence are that: (i) they almost never contain introns and
> thus the PCR product size is the same for cDNA and genomic DNA tem-
> plates; and (ii) the sequences of 3′UTRs are not as well-conserved as cod-
> ing sequences and thus it is much easier to distinguish between
> individual genes and paralogous gene family members that may be quite
> closely related in their coding sequences. ...One of the early problems
> with gene-based STSs was that there simply were not enough unique hu-
> man gene sequences to bother with. But all of that changed with the ad-
> vent of EST sequencing, at which time several groups began mapping
> ESTs albeit on a limited scale and usually only to the resolution of a chro-
> mosome assignment. The paper from Sikela's group in this issue reports
> the mapping of 318 cDNAs on the CEPH mega-YACs and is a milestone
> in the field, serving as a bridge between the limited EST mapping in the
> past and the thousands upon thousands of gene-based markers that will
> be appearing on maps in the near future. (Mark S. Boguski & Gregory D.
> Schuler; www.ncbi.nlm.nih.gov/Schuler/Papers/ESTtransmap/).

The identification of ESTs is a shortcut for identifying human genes since they are derived from active genes. ESTs can be obtained without any knowledge of their function. Since genes are not expressed all the time and often in a cell type-specific manner (as well as being dependent on different states of the development of an organism), the entire life cycle and all physiologically relevant tissues have to be probed for the presence of mRNA and its subsequent sequencing. This approach misses a major portion of an eukaryotic genome, but reveals interesting physiological and medical conditions (see functional genomics).

Non-coding DNA is sequenced fragment by fragment through PCR technology. The creation of the dbSTS depository includes specific sequence tags which can be used to uniquely identify chromosome locations (they serve as markers for genes when co-segregating). By using electronic PCR, STS with known chromosomal positions can be searched and compared with new sequences and the genomic position of the latter can be determined. In this way, e-PCR can be used for the creation of various types of genomic maps.

All in all, the rapid pace of sequencing often results in partial or unfinished sequences. The High-Throughput genome division of GenBank (http://www. ncbi.nlm.nih.gov/HTGS/) tries to accommodate this fact and coordinates submission of such fragments not only to GenBank, but also to the Japanese and European depositories. This is an effort coordinated among the three international nucleotide sequence databases: DDBJ, EMBL, and GenBank.

The Elimination of Redundancy

As the genome projects develop into an organized enterprise, the elimination of redundancy is a major concern in streamlining and optimizing the databases. Redundancy not only comes from the fact that different researchers are interested in the same protein or gene, but that different techniques of cloning and sequencing random genomes creates fragments with little biologically relevant annotation.

> One gene, many sequences. GenBank is a comprehensive source of sequence data, but selecting candidates for physical mapping can be difficult. This is in a large part due to the presence of multiple sequence records that, while not identical to one another, are derived from the same gene...Gene sequence entries possess differing amounts of flanking and intron sequence. Sequences of mRNAs can be incomplete or contain variation because of alternative splicing. Finally, ESTs are both fragmentary and have a higher error rate. For the UniGene set, all of these sequences are drawn together into a cluster if they are found to share statistically significant DNA sequence similarity in the 3'UTR. For the Washington University–Merck & Co. ESTs, these sequences are derived from oligo (dT)-primed mRNA, with directional cloning and sequencing from both the 5' and 3' ends. Clustering uses only 3' ends, but use of common clone identifiers places corresponding 5' ends into the clusters (Figure 1; www.ncbi.nlm.nih.gov/Schuler/Papers/ESTtransmap/).

Redundancy, of course, is beneficial for certain aspects of genome mapping and quality control, and redundancy and homology are closely related concepts with homology actually referring to two (or more) different genes with great similarity. Those are likely to be alleles in a population or the homologs of a specific gene in different species or taxonomic groups.

Human Gene Mapping Progress

It was projected in early 1998 that the human genome would be sequenced by the year 2005 based on the projected speed and number of base pairs in the entire genome. In October 1998, the new genomics company, Celera (http://www.celera.com/), formed by Perkin-Elmer (http://www.perkin-elmer.com/) and J. Craig Venter as a private initiative, and the U.S. government-funded Human Genome Project put the date of completion as early as 2001 with a "working draft" and a refined, accurate sequence map by 2003.

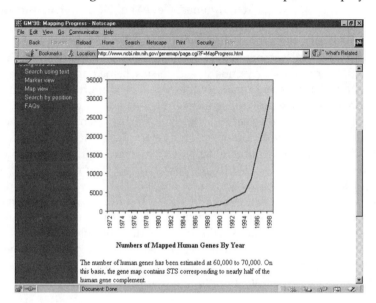

FIGURE 3.19
GeneMap '98; mapping progress

The implementation of new hardware and software were critical for this development. Behind the public effort stands the International Radiation Hybrid Mapping Consortium and the information is accessible through NCBI at http://www.ncbi.nlm.nih.gov/genemap98/. Major genome centers participating in the International RH Mapping Consortium are shown in the table below. For a complete list of participants refer to NCBI GeneMap '98.

TABLE 3.1

Major Genome Centers Participating in the International Radiation Hybrid Mapping Consortium

Genome Center	Country	Internet Address
Généthon	Evrey, France	www.genethon.fr/genethon_en.html
The Sanger Center	Cambridge, UK	www.sanger.ac.uk/
The Stanford Human Genome Center (SHGC)	Palo Alto, CA, U.S.	www-shgc.stanford.edu/
The Whitehead Institute for Biomedical Research (WICGR)	Cambridge, MA, U.S.	www.genome.wi.mit.edu/
The Wellcome Trust Centre for Human Genetics (WTCHG)	Oxford, UK	www.well.ox.ac.uk/

GeneMap '98 contained the positions of 30,261 genes on its release data (late 1998) showing an exponential growth in sequenced and/or positioned genes (Figure 3.19). The positions are marked by sequence-tagged sites (STS) and are unique (or appear to be unique) in an entire genome. Because of the small proportion of gene sequences in the human genome, only about 3% of all STS sequences correspond to actual genes. GeneMap '98 is derived form STS sequences that come from expressed sequences and represents this 3% of the genome. For this reason, the current map is also referred to as the *Human Transcript Map*. From the estimated 30,000 genes, about 15% can be found as full-length mRNA transcripts in GenBank with the rest being partial sequences only.

GeneMap '98 offers a summary of the gene distribution for every chromosome. Information regarding gene activity or gene dose effects can be obtained from this table because it compares the expected gene density with the observed density based on ESTs. It is obvious that gene density is not uniformly distributed along chromosomes, with the biggest deviations found in chromosomes 19 and 17 with higher density, and chromosomes 18 and X with markedly lower density than expected. The latter could be due to X-inactivation in females and males being XY. It is important to keep in mind that the current transcript map reflects gene expression activity rather than the presence of coding regions, with the gene density reflecting a functional distribution rather than a physical one. The possible differences in gene density obtained from ESTs and STSs mapping will show gene dosage effects, maternal factor inactivation, and parental imprinting.

3.4   Functional Genomics

With over 300,000 DNA sequences stored in databases around the world, the potential is tremendous for identifying "interesting" novel genes for medical and biological studies. It is evident from the twenty-something completed microbial genome projects that as many as 40% of structural genes are novel. This means that they have never been studied experimentally for their biochemical and physiological properties. Therefore, annotation of structure and function for those sequences which show some resemblance to known proteins relies on automated, statistical analysis. This method of predicting structure and function represents a crude, first run to gather biological information. These annotations, however, are increasingly based on previously predicted information. The biology behind the sequences, the phenotype–protein structure and function — still the systematic annotation of what is known biochemically about proteins — often falls behind. This means that even after the completion of a genome project, years of investigation are needed to understand the full complexity of the organism at the physiological level.

One of the first tasks after genomes are completely sequenced is to understand their content, i.e., to relate phenotype and genotype. In other words, the task ahead consists of assigning function to sequences based on the organization of genes within genomes and comparing these structures to distantly related genomes.

The task of extracting and analyzing the large amount of genomic data is made possible through public domain software that allows the analysis and characterization of all kinds of properties associated with DNA, RNA, and proteins. Tables 3.2 and 3.3 show the most common goals in studying genome structure, the identfication of novel genes and their related protein structures, and the major Internet addresses where these programs can be accessed.

TABLE 3.2

Public Domain Software Analysis Tools for DNA and RNA

Goal	Program	Internet Address
Sequence Similarity	BLASTn and tBLASTx and BLASTx	www.ncbi.nlm.nih.gov/BLAST
Finding Open Reading Frames (ORF)	ORF Finder	www.ncbi.nlm.nih.gov/gorf/gorf.html
Finding PCR-Based Sequence Tagged Sites (STSs) in DNA Sequence	Electronic PCR	www.ncbi.nlm.nih.gov/STS/
Translating DNA or RNA → Protein	Translate and Protein Machine	www.expasy.hcuge.ch/tools/dna.html and www.ebi.ac.uk/translate.html
Comparison of Genomic DNA and Protein Sequence	GeneWise	www.sanger.ac.uk/Software/Wise2/ genewiseform.shtml
Finding Genes	Gene Recognition and Assembly Internet Link (GRAIL) and PROCRUSTES	www.compbio.ornl.gov/Grail-1.3/ and www-hto.usc.edu/software/procrustes

The first step in understanding the relationship between genotype and phenotype is to look at the function of the entire genome. This is reflected in the cellular expression pattern of mRNA. Novel genes can thus be identified as being expressed in relation to a cellular activity for which we have some biologically significant information. The study of expression patterns then allows us to assign spatio-temporal information to unknown genes. Such a simple procedure can be used to structure a database according to perceived

TABLE 3.3

Public Domain Software Analysis Tools for Proteins

Goal	Program	Internet Address
Sequence Similarity	BLASTp and tBLASTn	www.ncbi.nlm.nih.gov/BLAST
Automated Structural Modeling	SWISS-MODEL	www.expasy.ch/swissmod/SWISS-MODEL.html
Identification and Characterization	Protein Identification and Characterization Programs	www.expasy.ch/tools/#proteome
Finding Patterns and Profiles	Pattern and Profile Search Programs	expasy.hcuge.ch/tools/#pattern
Structure Analysis	Primary Structure Analysis, Secondary Structure Prediction and Tertiary Structure Programs	www.expasy.ch/tools/#primary www.expasy.ch/tools/#secondary www.expasy.ch/tools/#tertiary
Sequence Alignment	Sequence Alignment Programs	/www.expasy.ch/tools/#align
2-D Page Analysis	Melanie II	www.expasy.ch/melanie/

correlations. Databases are subdivided to reflect different levels of function ascribed to proteins and genes. Hierarchical structures of databases are practical ways for biologists to quickly find information about a protein, gene, metabolic pathway, enzymatic activity, or evolutionary relationships. Current databases are constructed using information of related sequences, proteins, taxonomic information, predicted secondary structures, or domain organization of proteins.

It is also necessary to understand structure and function of protein families from organisms that are evolutionarily distant (i.e., they belong to different groups or kingdoms), like bacteria and humans. Sequence comparisons of distantly related organisms fall into three major classes. The first group comprises highly similar sequences and is observed for proteins of cell replication and information-storage processes. These proteins are true homologs, they share a common ancestral gene, and form a protein family. The second group includes proteins with similar structure and function, but dissimilar sequence. Their relationship can be inferred from structural and functional similarity, but lack significant sequence similarity. They may or may not be related evolutionarily and may be examples of convergent evolution. The third group shows no similarity for sequence, function, or structure.

Local sequence patterns that code for nucleotide binding pockets are valuable in assessing evolutionary relatedness because structural features of catalytic sites are better conserved than DNA or amino acid sequences of entire genes or proteins. This relationship conceptually allows us to search for conserved patterns in sequences and it is those patterns that are significant in establishing evolutionary relationships among genes. Even when the full-length (gene) sequence shows little or no similarity to other proteins, functional domains show a high degree of structural conservation. Because only selected stretches of fragments of genes show high similarity, these patterns can point out evolutionary mechanisms like gene duplication and recombination events leading to "chimeric" structures.

Adam Godzik of the Scripps Research Institute in La Jolla, California develops and applies novel algorithms to address some of the problems associated with identifying proteins with similar structure, but dissimilar sequence. The genome analysis page (http://cape6.scripps.edu/leszek/genome/) offers the comparison of sequence information of the genomes of *Mycoplasma genitalium*, *Escherichia coli*, and *Helicobacter pylori* with the protein structures in PDB. The program compares the predicted structures of all ORFs of an entire genome with all known crystal and NMR structures in the protein database. Comparing structural motifs allows us to identify weak relationships that are routinely missed by BLAST,[1] but still fails to predict the function for a large portion of bacterial genomes. As an example, the genome of *Escherichia coli* contains a total of ~4,300 genes coding for 1,500 hypothetical proteins (40% of all known genes or ORFs). From these, 30% (or ~500 ORFs) could not be predicted reliably as to what protein structure they encode or what the putative function of these proteins would be. An additional 30% could not be predicted at all, meaning they are completely novel proteins with no known counterparts in bacteria, archaea, plants, or animals.

Software used by Godzik and others implements *fold prediction* algorithms, profile/profile, structure/sequence, and structure/structure alignment (programs MODELLER[2]; COMPOSER[3]). These fold or local structure predictions make use of secondary structure prediction, buried amino acids, and contact patterns.

Unidentified Reading Frames: URFs

The genome projects completed thus far have identified approximately 30% to 40% completely new and unidentified gene sequences. They are referred to as URFs (unidentified reading frames) and no biological information is associated with them. No homologies are known, so they must be coding for new proteins that have not been found by biochemists or functionally identified by microbiologists. Structure prediction algorithms like that of Adam Godzik are of tremendous help here, but often do not help in understanding the function, indicating the big gap in knowledge of relating structure to function. Often it is a rationalization vaguely based on experimental evidence. Many structure prediction methods are *statistical* methods and rely on information obtained from known structures. The limited sample size (i.e., the number of known structures), limits the accuracy of predicting folds.

An alternative tool used to explore evolutionary relationship for novel genes is direct genomic comparison by studying the "behavior" of a genome. What is the mutation rate of known proteins of an organism? This information may help predict the uniqueness of a new gene in an organism that shows no sequence similarity to any known protein. Assuming that the mutation rate is even in the entire genome of an organism (an assumption which is not necessarily true), the sequence dissimilarity of a URF sequence is an indicator that it may belong to a new class of proteins that define the uniqueness of the organism.

Take as an example the enzymes that link amino acids with a small class of RNA molecules called transfer- or tRNA (see F. Doolittle, 1998, *Nature* 392,339). In all mammalian organisms there are at least 20 different enzymes, one each for each of the 20 amino acids used to synthesize proteins (a universal fact of life). From the completed genome project of the archaea *M. jannaschii*, for four of these *amino acyl tRNA synthetase* no corresponding gene has been identified, although they must exist, since all tRNAs are linked properly with their respective amino acids.

One possibility explaining the lack of genes is to assume a totally new mechanism for amino acyl tRNA synthesis: chemical modification of amino acids linked to tRNA molecules. One of the unidentified genes codes for the *lysyl-tRNA synthetase*. However, it has been demonstrated by functional cloning in a research project unrelated to the genome project, that this synthetase responsible for linking the amino acid lysine to its tRNA partner has a sequence totally unrelated to any known lysine-tRNA syntheses. Indeed, this is an example of an *entirely new family of proteins*. The conclusion is that two entirely unrelated proteins, as judged on the bases of their DNA sequence, perform the same enzymatic activity. This has been shown for other classes of enzymes such as the serine protease chymotrypsin and subtilisin. They catalyze the same chemical reaction using structurally conserved active sites, albeit with different substrate specificity.

The existence of evolutionarily unrelated but functionally similar proteins demonstrates that in the complete absence of functional data, the interpretation of what kind of protein DNA sequences code for can be difficult and sometimes impossible. The use of data from functional cloning efforts was necessary to assign a biological function to URFs. To understand the relationship between structure (sequence) and function of the amino acyl tRNA synthetases requires a great deal of familiarity with the topic. Scientists unfamiliar with tRNA metabolism are not likely to find the apparent relationship.

Cluster of Orthologous Groups: COGs

Finding functional relationships between genes across taxonomic groups (orthologs) as well as within a population or the same organism (paralogs) provides the true potential of genome projects. Comparing protein sequences encoded in eight complete genomes, representing six major phylogenetic lineages (http://www.ncbi.nlm.nih.gov/COG/; April 1999) delineated clusters of orthologous groups (COGs; Figure 3.21). This is an effort to use databases to generate new information by linking sequence information from various complete genomes. Any two proteins from different lineages that belong to the same COG are orthologs according to the functional definition used by NCBI to construct COGs, and are assumed to have evolved through speciation. COGs also contain paralogs, which arise through gene duplication events.

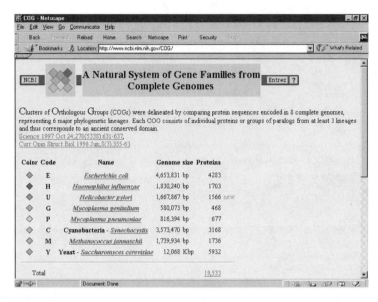

FIGURE 3.20
COG: A natural system of gene families from complete genomes

Using the completed genomes of eight organisms (*E. coli; H. influenzae; M. genitalium; H. pylori; M. pneumoniae; Synechocystis; M. jannaschii; S. cerevisiae*), a total of 864 COGs have been identified belonging to information storage and processing (groups J, K, L), cellular processes (groups O, M, N, P), metabolism (groups C, G, E, F, H, I), and predicted or unknown functions (groups R, S). This group of poorly defined proteins comprises a total of 180 COGs (out of 864) with 1828 proteins and domains associated to predicted function (R), and 271 uncharacterized proteins and domains (S). The analysis of COGs allows for the understanding of evolutionary relationships and the identification of related functions across taxonomic divisions.

Group N contains 20 COGs, one of which represents the signal peptidase family of proteases involved in protein secretion in eubacteria and eukaryotes, but not archaea (COG ID 0681; Figure 3.22). Signal peptidase I is a small, membrane-bound protein that cleaves off the N-terminal signal sequence of proteins transported across the endoplasmatic reticulum membrane of eukaryotes and inner membrane of bacteria and mitochondria (the current members of completed genomes include seven bacterial and one eukaryotic

Bioinformatics Basics

organisms). Signal peptidase I COG consists of eight members, one *E. coli* protein LepB, the *H. influenzae* protein HIN1152, *Synechocystis sp.* paralogs slr1377 and sll0716, *M. jannaschii* protein MJ0260, and three yeast paralogs, proteins YMR150c, YMR035w, and YIR022w.

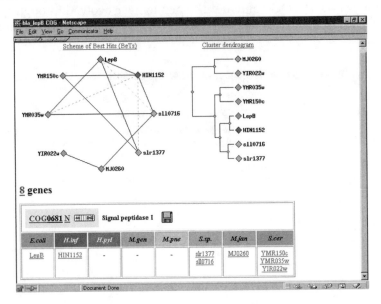

FIGURE 3.21
COG0681 for signal peptidase I

The COG demonstrates the complexity of searching for phylogenetic relationships among proteins of distantly related organisms. One of the yeast paralogs (YMR035w) shows best-hit similarity to three different bacterial species (*E. coli, H. influenzae,* and *Synechocystis* sp.). YMR150c and YMR035w are inner-membrane proteases of yeast mitochondria, responsible for removal of signal peptides from some proteins of the inter-membrane space, but with different substrate specificity. YIR022w is the yeast signal sequence processing protein in the endoplasmatic reticulum, required for signal peptide cleavage and normal rate of protein secretion.

The cluster dendrogram shows the close relationship of mitochondrial proteases with bacterial orthologs and the relationship of the ER protease paralog with the archaea ortholog of *M. jannaschii*. This is consistent with the proposed common ancestral single-cell organism of eubacteria and mitochondria (endosymbiotic theory). That the *M. jannaschii* protease shows an orthologous relationship to both yeast and cyanobacteria, stresses the taxonomic classification of archaea as different from both eukaryotes and eubacteria. This COG analysis indicates that of the three yeast peptidases, the two mitochondrial subtypes are true paralogs with a common eubacterial origin, whereas the ER subtype evolved independently or originates from an older ancestral gene that precedes split into eubacteria and archaea. This split may be the event behind the paralogs in *Synechocystis* sp.

The COG database lists related genes in patterns indicating their occurrence in different organisms (Figure 3.23). One such pattern reads *eh−cmy*, which excludes the two Gram-positive, pathogenic mycoplasmodia and the ulcer-inducing *H. pylori*, but includes the Gram-negative pathogen *H. influenzae*. Thirty-nine other COGs with this pattern have been identified, including the one for signal peptidase I. The COG which represents the most commonly found phylogenetic pattern, including all eight genomes in this analysis, contains 110 clusters, most of which belong to functional group J relating to translation, ribosomal structure, and biogenesis. Other groups of enzymes belong to such central metabolic pathways as glycolysis, pentose phosphate pathway, RNA polymerases, protein folding, and secretion.

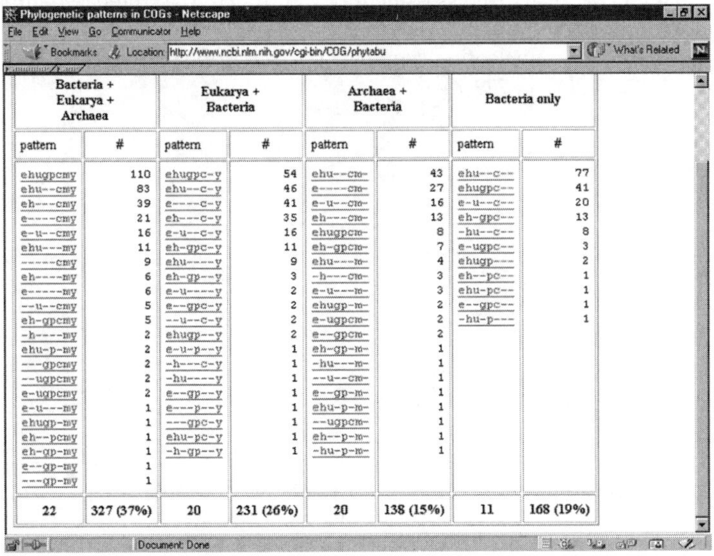

Bacteria + Eukarya + Archaea		Eukarya + Bacteria		Archaea + Bacteria		Bacteria only	
pattern	#	pattern	#	pattern	#	pattern	#
ehugpcmy	110	ehugpc-y	54	ehu--cмo-	43	ehu--c--	77
ehu--cmy	83	ehu--c-y	46	e-----cмo-	27	ehugpc--	41
eh---cmy	39	e----c-y	41	e-u--cмo-	16	e-u--c--	20
e----cmy	21	eh---c-y	35	eh----cмo-	13	eh-gpc--	13
e-u--cmy	16	e-u--c-y	16	ehugpcмo-	8	-hu--c--	8
ehu---my	11	eh-gpc-y	11	eh-gpcмo-	7	e-ugpc--	3
-----cmy	9	ehu----y	9	ehu---мo-	4	ehugp---	2
eh----my	6	eh-gp--y	3	-h----cмo-	3	eh---pc--	1
e-----my	6	e-u----y	2	e-u---мo-	3	ehu-pc--	1
--u--cmy	5	--u--c-y	2	ehugp-m-	2	e---gpc--	1
eh-gpcmy	5	ehugp--y	2	e-ugpcмo-	2	-hu-p---	1
-h----my	2	e-u-p--y	1	e---gpcмo-	2		
ehu-p-my	2	-h---c-y	1	eh-gp-m-	1		
---gpcmy	2	-hu----y	1	-hu---m-	1		
--ugpcmy	2	e---p--y	1	e---gp-m-	1		
e-ugpcmy	2	---gpc-y	1	ehu-p-m-	1		
e-u---my	1	ehu-pc-y	1	eh--p-m-	1		
ehugp-my	1	-h-gp--y	1	-hu-p-m-	1		
eh--pcmy	1						
eh-gp-my	1						
e--gp-my	1						
---gp-my	1						
22	327 (37%)	20	231 (26%)	20	138 (15%)	11	168 (19%)

FIGURE 3.22
Phylogenetic patterns in COGs

The importance of identifying such patterns where different organisms share sets of enzymes or pathways can bring about information on the biochemical requirements necessary for survival in different environments.

The group of pathogens well suited for study of the minimal genetic requirements for host infection and replication is viruses. Viruses are less complex systems because they make use of the cellular machinery of the host organism. Their genome appears to have adapted and streamlined presence of necessary genes. Viruses are the best adapted, but so are host-dependent minimal organisms containing minimal genomes. Because of their small size, viral genomes have been sequenced long before that of the first pathogenic bacteria *H. influenzae*. The DNA sequence for the genome of bacteriophage ΦX174 with 48,502 base pairs is the first viral genome sequenced by Frederick Sanger's group in 1982[4]. *H. influenzae's* 1.7 million base pairs were later sequenced in 1995.

―――――――

References

1. Altschul, S.F., et al., Basic local alignment search tool. *J. Mol. Biol.*, 1990. 215(3): p. 403-10.
2. Sali, A. and J.P. Overington, Derivation of rules for comparative protein modeling from a database of protein structure alignments. *Protein Sci.*, 1994. 3(9): p. 1582-96.
3. Topham, C.M., et al., An assessment of COMPOSER: a rule-based approach to modelling protein structure. *Biochem. Soc. Symp.*, 1990. 57: p. 1-9.
4. Sanger, F., et al., Nucleotide sequence of bacteriophage lambda DNA. *J. Mol. Biol.*, 1982. 162(4): p. 729-73.

4

Proteome Analysis

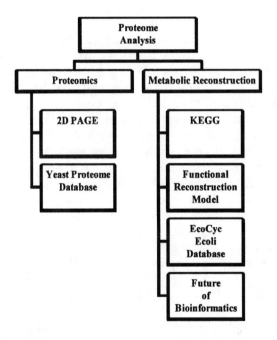

FIGURE 4.1
Chapter overview

4.1 Proteomics

Most databases are built from sequences of genes, genomes, and proteins. However, little is understood about how macromolecules — nucleic acids, proteins, lipids, and carbohydrates — work together to give structure and function to a cell, be they single-cell organisms like bacteria and yeasts, or multicellular organisms like plants and animals. Furthermore, because mRNA is the intermediate molecular species for protein biosynthesis, mRNA levels are indicative of gene expression. For this reason, mRNA sequences are used for the production of EST (expressed sequence tags) libraries. However, cellular levels of mRNA are not necessarily reliable markers for protein levels. It is therefore of paramount importance to establish the protein profile of a cell, which not only gives the relative number of proteins, but also the form in which they exist, and posttranslational modification such as glycosylation, acylation, ubiquitination, phosphorylation, or proteolytic processing involved in protein activation. All these processes are involved in controlling the activity and location of a protein in the cell.

To make matters more complex, the protein composition of a cell and the posttranslational modifications of proteins vary at different stages of the cell cycle, in metabolic and environmental stress, in cell-to-cell signaling, and in individuals with diseases. Tumors, for example, often show altered expression and activity patterns of key proteins as compared to healthy tissue. Since these proteins are related to growth control, carcinogenesis can generally be viewed as some lack of control in growth, or the unhindered multiplication of cells which should no longer divide or are programmed to die (aging, apoptosis).

Because all somatic cells contain the full genome, but use only part of it for regular activity, up and down regulation of genes is especially important during the development of these cells, where a single cell multiplies and differentiates into specialized cell types, tissues, and organs. The imprinting of gene activity patterns is a process where the cells control the levels of gene products and behave as recessive or dominant genes. This is known as gene dose effect and is well studied for X chromosome-related genes that contain one copy in males and two copies in females. The level of proteins often affects metabolic and second messenger pathways and is used as a fine-tuning control mechanism. The cellular mechanisms of such gene dose effects are often not understood.

Furthermore, not all macromolecules and cellular structures are synthesized off a linear template. In fact, RNA and protein synthesis are the only molecular species directly encoded for by DNA. Everything else — and this includes the protein modifications mentioned earlier — is guided by molecular interactions, sequential synthesis, and spatial separation known as compartmentalization. Good examples of biological macromolecules lacking a gene template are *polysaccharides* and protein and lipid glycosylation. Polysaccharides (carbohydrates) exist as linear, as well as branched, multimers, and although polymer sequences are consistently reproduced by the cellular machinery (proteins catalyze carbohydrate synthesis), these sequences are not encoded by other linear, molecular templates as DNA codes for protein synthesis. Instead, polysaccharide synthesis is a sequential catalytic activity performed by the spatial arrangement of enzymes within the cell. Its own gene encodes each enzyme of such a pathway. Groups of enzymes that synthesize polysaccharides are therefore not independent, since the lack of a single protein in the pathway makes it defective.

It is therefore important to understand the structure of enzymatic pathways. Comparing not only individual genes across species, but entire pathways yields additional information about newly discovered DNA sequences. Are pathways identical across species? Are all enzymes of the same pathway homologs expressing similar degrees of identity? Are certain enzymes in pathways more important or more conserved than others? Do some species have alternative pathways to generic ones, while others do not? Finding the answers to such questions is the true challenge of biology in the twenty-first century. The Internet (or any equivalent form of public communication) will be instrumental in this discovery process. It will provide the databases necessary for comparing the protein composition of a cell or an organism as a function of metabolic activity and disease from the period of conception to the moment of death.

The task of organizing the collection of proteins of any given cell type or organism has recently given rise to a new science called *proteomics*, which attempts to look at the combination and expression patterns of a cell's or an organism's proteins at any given moment. It basically tries to understand the organizational complexity of the enzymatic machinery of cells. The word proteomics refers to the idea that all proteins of any given organism are necessarily linked in their fate with each other. Understanding this interrelationship is important in understanding their biology, including their evolutionary traits (see also codon bias in yeast, mitochondria).

Proteomics makes use of a biochemical technique invented in the early 1970s (O. Farrell and J. Klose) where proteins are analytically separated on polymer gel electrophoresis in two dimensions using molecular weight for the first dimension and electrical charge as a function of pH for the second dimension (Figure 4.2). The so-called two-dimensional gel electrophoresis (2-D gels) can be used for comparative purposes to identify proteins that exist in different quantities, modification, and different times during the life cycle of an organism. It is important to compare sets of proteins to one another, not only to replace the more tedious work of studying single proteins, but also to be able to correlate coexpression of proteins and to relate these patterns to cellular activity.

Proteomics is a formidable approach because many proteins have never been characterized on 2-D gels (or characterized at all) and a biochemist's most time-consuming work is to unambiguously identify the spot on an analytical gel as a specific protein, modified protein, or fragment thereof. As previously mentioned, 2-D gels give two pieces of information: size and charge. Both physical parameters depend on the cellular condition at the time the protein was isolated and purified. The calculated molecular weight of a protein based on its DNA sequence often does not exactly match the experimentally determined value from gel electrophoresis, because the latter reflects the overall solubility and mobility of a protein within the gel matrix. The mobility in the direction of molecular weight not only depends on the true molecular weight of the protein, but more accurately reflects the charge/unit weight of the protein. Therefore, changing the pH of the system changes the mobility because the number of charges-per-unit weight is changed and not

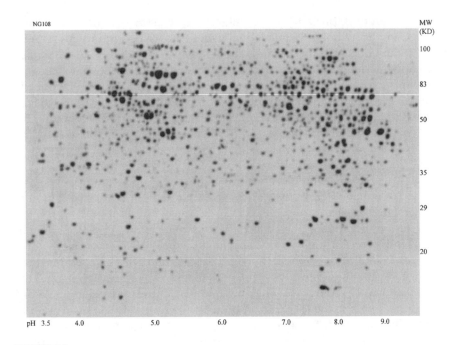

FIGURE 4.2
2-D polyacrylamide gel showing protein contents of eukaryotic cell type: proteins extracted from the mammalian cell line NG108 are separated according to molecular weight (MW in kilodalton (KD)) on Y-axis and according to charge in a pH gradient ranging from 3.5 to 10 (X-axis). Each spot represents an individual protein type. Intensities reflect protein concentrations. Proteins can be extracted from gel matrix for biochemical analysis (sequencing, mass spectrometry). (with permission, Young Yang, R.W. Johnson Pharmaceutical Research Insitute, San Diego, California).

every protein with the same molecular weight has the same charge/unit weight ratio. The precise determination of the actual molecular weight and sequence of a protein spot on a gel is crucial for such interpretation.

Modern analytical and automated systems assist in the large-scale identification of these protein spots (Figure 4.3). Proteins of interest can be digested within the gel matrix and the resulting peptides are extracted and subjected to high mass accuracy MALDI-MS (matrix-assisted desorption ionization mass spectrometry) analysis. Here, peptide fragments are ionized and their charge/mass ratio is determined. The mass-charge ratio is matched to all possible amino acid sequence combinations. If the matching is ambiguous, the peptide fragment must be micro-sequenced and the sequence subjected to a database search using BLAST algorithms. If many fragments from a single 2-D gel spot match the same sequence in the database (e.g., GenBank), the protein corresponding to this spot has been successfully identified.

The matching is often not straightforward because of the potential chemical modification of the peptide fragments. These posttranslational modifications come from cellular processes used to control protein activities. These

Strategies for the Identification of Proteins from 2-D Gels

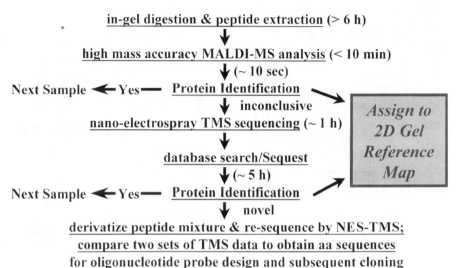

FIGURE 4.3
Strategies for the identification of proteins from 2-D gels: see text for details (with permission, Young Yang, R.W. Johnson Pharmaceutical Research Institute, San Diego, California).

modifications affect net charge, reactivity, and solubility of proteins. Modifications such as phosphorylation add negative charges to the protein, thus influencing its mobility during electrophoresis. A single negative charge has the equivalent effect of decreasing the molecular weight of a protein by 2 kDa, or roughly 15 to 18 (non-charged) amino acids. Glycosylation also affects the molecular weight of a protein, but not necessarily its pH dependence. Because of the existence of multiple modifications that affect the apparent mobility of a protein on a gel in a similar way, the interpretation of small differences in mobility of proteins on 2-D gels is not always easy and requires careful biochemical analysis.

The entire process of peptide fragment identification has been automated over the last several years (Figure 4.4). Automated processes require special robotic equipment, as well as customized software. Again, computers play a central role in controlling and analyzing the process. An autosampler collects peptide fractions from a HPLC column chromatography which separates peptides according to size. Very small volumes are used in capillary columns and subjected to nano electrospray ionization for mass spectrum analysis. Experimental and predicted mass spectra are used to generate cross-correlation data to identify the sequence of the extracted peptide fragments. If sev-

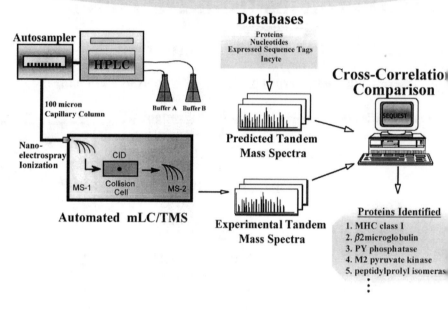

FIGURE 4.4
Fully automated protein identification (with permission, Young Yang, R.W. Johnson Pharmaceutical Research Institute, San Diego, California)

eral fragments from a single 2-D gel spot match a single amino acid sequence entry in the database, a protein is identified.

2-D Page at Expasy

The first step in proteomic analysis of cellular mechanisms is to compare 2-D gels of cellular extracts obtained after stimulating a cell with an activator (such as insulin on liver cells) with those obtained under metabolic resting conditions.[1] Many public databases include a growing number of such reference gels for preliminary identification of the charge and molecular weights of a novel protein. A public proteome database (SWISS-2DPAGE) has been established at the Geneva University Hospital, Switzerland (http://www.expasy.ch). This database is one of eight intended to help scientists understand organisms at the functional level by directly studying gene products (proteins) and their corresponding posttranslational modifications. The site is organized to access the 2-D database interactively, to provide online help, technical manuals for 2-D gel electrophoresis, and services such as running gels upon submission of samples, training courses (not online courses, you have to travel to Geneva), and software packages for the analysis of 2-D gels.

The 2-D database at Expasy contains electrophoretic information on 16 tissues and organisms, including yeast, *E. coli, Dictyostelium,* and human cell types (platelet, red blood cell, macrophage, plasma protein, lymphoma, liver, kidney, two leukemia cell lines, cerebrospinal fluid, and intestinal epithelial cells). Known proteins can be found by searching for an accession number (SWISS-PROT) or clicking on 2-D gels with marked spots. The putative locations of new proteins can be identified if the amino acid sequence is known. The hypothetical molecular weight and charge are used to place the protein on the gel. However, and this is an important caveat for 2-D gel electrophoresis, theoretical values often differ from experimental values due to solubility behaviors of proteins in gels and actual posttranslational modification of the amino acids. Many proteins yield multiple spots on 2-D gels for this reason, so this information is valuable to the biochemist in understanding the function of a protein within a cellular environment.

A look at those gels shows that the majority of spots are not linked to any known protein. New technology is being developed to more quickly identify proteins on 2-D gels. Biochemical analysis using microsequencing of peptide fragments and mass spectrometry of these fragments is the analog approach in sequencing nucleic acid libraries.

Once a protein is identified in one cell type or organism, its expression level can be compared in other cell types or tissues, potentially revealing different levels of expression and modes of posttranslational modifications. This task of comparing protein expression levels is hardly a trivial one. The way a protein runs on a gel greatly depends on the purification procedure, source, and electrophoresis procedure. Comparison, therefore, requires cautious interpretation regarding relative positioning of spots and their intensities. SWISS-2DPAGE offers an analysis package for rapid image manipulation, complete 2-D analysis, worldwide comparison for referencing, and automated gel matching and comparison (Melanie II 2-D Analysis Software, developed by Denis Hochstrasser at Melanie Group in Geneva, http://www.expasy.ch/melanie/MelanieII/description.html). The features of Melanie II include:

Rapid Image Manipulation:

- Zooming
- Filtering (smoothing, contrast enhancement, background subtraction)
- Gel flipping
- Gel stacking for better visualization
- Image stretching

Complete 2-D Analysis:

- Automatic spot identification and analysis

- Gaussian spot modeling
- Gel overlay display
- Point-and-click interface
- Embedded landmarks
- pI/MW setting
- Extended reports
- Histograms
- Statistical data analysis

Worldwide Comparisons:

- Multiple gel display
- Fast, automatic gel comparison and matching
- Reference gels for comparison to all other gels
- Creation of synthetic gels by merging a set of gels
- SWISS-2DPAGE master gels
- Network and online links to biological databases, including SWISS-2DPAGE and SWISS-PROT through Expasy
- World Wide Web server

Data Import/Export:

- Gel printing
- Image import/export from and to TIFF and PPM
- Data export to Excel and other applications
- Data export as Melanie I format to public statistical and heuristic clustering programs.

Specialized 2-D Page Databases

There are two specialized proteomic databases that compare expression profiles associated with *toxins* and *xenobiotics*: the Rodent Molecular Effects Database at Oxford's Glycosciences, and a keratinocyte database at the Danish Centre for Human Genome Research at http://biobase.dk/cgi-bin/celis. The latter provides a database for knockout and transgenic animals, meaning that specific genes have been inactivated or added to the germ lines of the animals. The absence or addition of genes should be observable at the protein level by a missing or additional spot on a 2-D gel. ESA, Inc.'s neurological disease database specializes in Alzheimer's, Parkinson's, and Huntington's diseases through protein differential display.

The Yeast Proteome Database™ (YPD) from Proteome, Inc. (www.proteome.com) is an example of how entrepreneurial efforts tap into a vast amount of existing data in the current literature and combine it into a specific new form; in this case, the collective knowledge of all proteins of the microorganism *Saccharomyces cerevisiae*, baker's yeast (genome project completed in 1997).

YPD[2] is an encyclopedia of all yeast proteins known and predicted from the yeast genome project connecting basic biophysical and functional data like molecular weight from mass spectrometry, amino acid sequences (from genome sequences), and function (from published literature). Currently, about 30 new yeast proteins are being characterized at different levels of information, and 3000 proteins are known to some degree (how many ORFs and URFs). In addition, homology-based information that cross-references yeast with human, for example, is valuable to researchers who may use yeast as a model organism to perform initial studies on proteins that are used similarly in human metabolism and physiology. The data going into YPD and 2-D gel maps in general include molecular weight information gained from *mass spectrometry*, the charge and chemical modification information gained from the *amino acid sequence*, and *functional* information largely gained from published data.

References

1. Wilkins, M.R., et al., Protein identification with N- and C-terminal sequence tags in proteome projects. *J. Mol. Biol.*, 1998. 278(3): p. 599-608.
2. Garrels, J.I., YPD-A database for the proteins of Saccharomyces cerevisiae. *Nucleic Acids Res.*, 1996. 24(1): p. 46-9.

4.2   Metabolic Reconstruction

Kyoto Encyclopedia of Genes and Genomes — KEGG

The Kyoto Encyclopedia of Genes and Genomes (KEGG) is an effort to provide a means to compute pathways of molecular and cellular processes. KEGG[1] is part of the Japanese Human Genome Program at the Institute for Chemical Research, Kyoto University (http://www.genome.ad.jp:80/kegg/; also see Chapter 2.1 for details on using KEGG). Although the technical challenges and underpinnings of KEGG are the same as those of NCBI, KEGG addresses the complex interaction of proteins in cells from the perspective of metabolic interaction. Their philosophy includes finding answers to some of the common questions of modern molecular biology such as: what do we know about the relationship between the sequence of a gene and the function of the protein? What is the protein-folding problem in the *cellular* context? What are the challenges and problems facing the functional reconstruction problems, or how can we understand the relationship between the genome and the organism — its development and morphology? The goal of KEGG, therefore, is to build a functional map starting with available components within various molecular catalogs. The functional maps represent metabolic and regulatory pathways. The molecular catalogs include genome maps, base sequences, gene allocations (physical map, inheritance map), and LIGAND databases (enzymes, compounds, and elements).

The Functional Reconstruction Model

Other databases and organizations have information similar to KEGG's, but with a notable exception: KEGG also has a *deductive database*. Here, the user can compute pathways and binary relations with the goal of computing wiring diagrams of genes and molecules. This should lead to an understanding of cells as being a complex, self-assembling system whose components and the relationship between them are fully understood. This is known as the functional reconstruction model. In KEGG's words, the genome is simply a "warehouse of parts...and all the regulatory signals in the genome are simply bar codes to retrieve them. In this view the blueprint of life is written in the entire cell as a network of molecular interactions." To gain an understanding of this network of molecular interactions, KEGG uses *prediction tools* for the computation of novel relationships based solely on the contents of its individual catalogs (the warehouse). These tools can be found through the "Search and compute with KEGG" link.

How can we reconstruct a biological organism? KEGG's approach uses a hierarchical view of an organism, most easily viewed as atomic levels, molecular levels, and network levels (pathways). KEGG uses a system for data representation that structures a database according to the number of links between its components.

Elements of the catalog database include molecules (protein structures, metabolites), genes (sequences), and genomes (Figure 4.5). Pathway maps connect elements from the catalog database through binary relations, which are molecular interactions (structure) and genetic interactions (function). Interactions of more than two units are called networks and include metabolic pathways (molecular and genetic), genomes (linear and circular), hierarchies (classification, taxonomy), and neighbors (sequence similarities, structural similarities).

KEGG introduces the integration of genomic information with pathways to reflect the biological reality within a cell. This allows the individual scientist

Database hierarchy

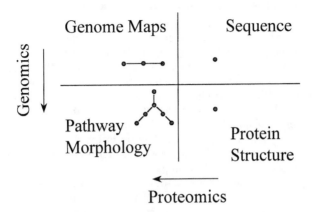

FIGURE 4.5
Database hierarchy

to search for proteins or genes of new or related pathways in model organisms other than the one being investigated. Missing structural information can quickly be obtained by finding homologs in "neighbors" whose structure has been solved. Novel pathways may be predicted by entering starting and end points of substrate and product and selecting an appropriate organism.

E. coli Metabolic Database: EcoCyc

The bacterium *E. coli* is the ultimate laboratory test subject for geneticists, molecular biologists, microbiologists, and biochemists. It is also extremely important to human health and physiology, as it is part of our gastrointestinal system. Unfortunately, *E. coli* is also an opportunistic pathogen; i.e., it can cause lethal infections if it enters our bloodstream. It is recognized by most people for its role in food poisoning caused by meat contamination, most often in undercooked hamburger meat. Its close genetic relationship to the bacteria *Salmonella typhimurium* (a problem found mostly in poultry) makes the integrative knowledge of its metabolism, genetics, and health problems a pressing yet fascinating issue.

EcoCyc[2] from Pangea Systems, Inc. (http://www.pangeasystems.com/) addresses the integration of the classic biochemical pathways (metabolism) of *E. coli* with its complete genome sequence information. For example, the metabolic pathways for amino acid synthesis in *E. coli* involve several enzymes that are often coregulated at gene expression levels. Thus, sensitive proteomics techniques should be able to see shifts in spot density, not only for one protein, but for several different ones. New pathways or homologous pathways in recently studied organisms with limited sequence information may be detected by proteomic means.

Similar to KEGG, EcoCyc uses chemical compound libraries that list the molecules involved in each biological reaction, the molecular weight of the compound and, in many cases, its chemical structure.

> The EcoCyc KB has a number of uses. It is an electronic reference source for *E. coli* biologists and for biologists who work with related microorganisms. Scientists can visualize the layout of genes within the *E. coli* chromosome, or of an individual biochemical reaction, or of a complete biochemical pathway (with compound structures displayed). The navigation capabilities allow the user to move from a display of an enzyme to a display of a reaction that the enzyme catalyzes, or of the gene that encodes the enzyme. The interface also supports a variety of queries, such as generating a display of the map positions of all genes that code for enzymes within a given biochemical pathway.
>
> In addition to being a reference source for individual facts, the EcoCyc KB allows complex computations related to the metabolism, such as design of novel biochemical pathways for biotechnology, studies of the evolution of metabolic pathways, and simulation of metabolic pathways. The EcoCyc KB is also being used for computer-based education in biochemistry. (From: http://ecocyc.PangeaSystems.com/ecocyc/ecocyc.html)

The *E. coli* metabolic database nevertheless provides us with a considerable body of knowledge that is easily accessible from a PC terminal. As of January 1998, the following numbers of objects were contained in the latest version of EcoCyc:

- 4909 *E. coli* genes
- 804 enzymes encoded by these genes
- 829 metabolic reactions occurring in *E. coli*
- 124 metabolic pathways occurring in *E. coli*
- 1303 chemical compounds involved in *E. coli* metabolism
- 79 tRNAs
- 45 2-component signal transduction proteins

What keeps the bacterium alive? This question looms in the minds of life scientists and the answer seems frighteningly close. The creators of EcoCyc talk about "an *in silicio* model of *E. coli* metabolism that can be probed and analyzed through computational means."[3] This suggests that experimenting on the model of metabolic pathways instead of the biochemical (wet bench) model will someday be as possible as a computer simulating nuclear tests thus replacing the actual tests which are detectable seismographically by the enemy. Further, it suggests that the encyclopedia will someday be transformed into a workbench — an electronic laboratory rather than an electronic library.

Where Should We Be in the Future?

At the current level of database integration and the combination of general information, the Internet is an open-ended, badly annotated, and unedited "database." There are infinite numbers of existing hyperlinks, and the task of searching for correct and complete information is often difficult and sometimes impossible.[4] Small companies are often founded on the realization that editorial work, annotation, and human intervention are needed to enable the life science community to make full use of the existing information. Annotation is costly, but can sell well if it is customized to the needs of the end user. For entrepreneurs, selling compounded information to the industry can be very lucrative, since the time and expertise required for scientists to work together with computer engineers can be financially prohibitive. Selling information over the Internet — information that is already available in unprocessed form — is vulnerable to exploitation. The interface between proprietary and academic information and the struggle for new patents on life forms feeds companies such as Proteome and Pangea Systems.

Spatial organization of cellular components is central to the function of a cell, yet this aspect of supramolecular structures is poorly understood. It boils down to how cells build the necessary building blocks at the right place and the right time. Self-assembly systems are part of a new interdisciplinary branch of biology and chemistry that addresses such problems — chemistry beyond the molecule.

References

1. Ogata, H., et al., KEGG: Kyoto Encyclopedia of Genes and Genomes. *Nucleic Acids Res.*, 1999. 27(1): p. 29-34.
2. Karp, P.D., et al., Eco Cyc: encyclopedia of *Escherichia coli* genes and metabolism. *Nucleic Acids Res.*, 1999. 27(1): p. 55-8.
3. Karp, P.D., et al., EcoCyc: enyclopedia of *Escherichia coli* genes and metabolism. *Nucleic Acids Res.*, 1997. 25(1): p. 43-51.
4. Lawrence, S. and C.L. Giles, Searching the World Wide Web [in-process citation]. *Science*, 1998. 280(5360): p. 98-100.

5

The Computer Revolution in Neurobiology

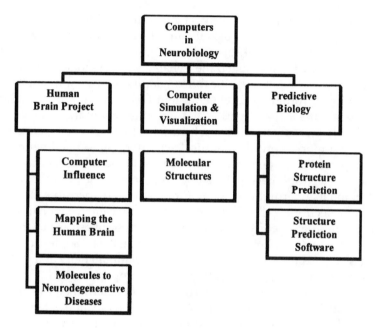

FIGURE 5.1
Chapter overview

5.1 Human Brain Project

Computers are used in almost every modern lab's equipment including power supplies for gel electrophoresis, computer-controlled chromatography, electrophysiology for the study of action potentials (ion channel, receptors for neurotransmitters, hormones), spectroscopic techniques (fluorescence, NMR, crystallography, electron microscopy), online recording (replacing tape recordings), online oscilloscopes (simulating analog machines), and digital oscilloscopes with computer access.

The Influence of Computers in Science

Experimental science in biology, physics, and chemistry conjures up images of test tubes, colored liquids, microscopes, and men in lab coats with wild hair. Yet daily life in the laboratory can be as banal to a scientist as it is mysterious to the average person. We know scientists experiment, but what does that mean? Experimenting is the testing of an idea by organizing a set of chemicals over a specified time period. What happens next is not left to chance, but expected by necessity — the necessity of the observer's assumptions and physical parameters, such as temperature, acidity, concentration, or pressure. Experimenting obviously requires planning and organization.

Computers play a central role in the design, execution, and analysis of experiments. They have changed the dynamics of the modern laboratory in almost every respect, not scientific thinking or methods, nor hypotheses and controversies surrounding discoveries and falsification of models, but the core of science: the actual experiment. Computers count the number of cells in a petri dish, measure the size of cell nuclei in the microscopic cross section of a kidney tumor, record the electrical activity of an isolated neuronal cell that is suspected in chronic pain sensation, and read the sequence information from an electrophoresis gel after having exposed the radioactive labeled DNA fragments overnight on X-ray film. Other indispensible lab support comes from electronic cell sorting machines, video cameras with online control from computers, purifying a product in reverse-phase chromatography, digital oscilloscope, and real-time peak current analysis. Sophisticated computers assist in experimental systems where the human hand and eye would be too slow and inaccurate to guide and record important data.

The precision of an experiment, of course, is not determined by computers, but by the quality of the instrument used. High-quality alloys must be engineered for the precision cutting of frozen cell samples in electron microscopy, the pulling of a micro-glass pipette for the transfer of a cell nucleus into stem cells to produce knock-out mice, or measuring the electrical activity of a single neuron in brain tissue. The computer's functions are for control, performance, data manipulation, and storage. Storage capacity has increased dramatically due to the computer's digitalization of recordings and ease of reproduction.

It must be remembered, however, that precision and high-quality measurements in science were achieved long before the advent of the computer. Ramon y Cajal (1852–1934), a Spanish neuroanatomist and the 1906 Nobel laureate in medicine, exemplifies the scientific excellence of the nineteenth century (biographical sketch at http://www.nobel.se/essays/cajal/index.html). He studied the nervous system and recorded his anatomical studies of the brain in meticulous, accurate drawings (Portrait of Cajal, http://www.faseb.org/anatomy/bict5.gif).

Cajal was the leader in brain research because of his state-of-the art drawings of neurons and groups of neurons based on new staining techniques (dye coloring individual cells or groups of cells and showing the interconnectedness of neurons) developed by Camillo Golgi, the co-winner of the 1906 Nobel prize in medicine. Cajal showed that the vertebrate brain was made up of billions of individual cells and neurons and was not a continuous network of fine arteries. Similarly, new staining techniques combined with photography and computer-aided spectroscopy are used today to construct anatomical maps of the human brain (see also "the whole brain Atlas" at Harvard Medical School; http://www.med.harvard.edu/AANLIB/home.html). Staining and sample preparation that make use of a variety of molecules are still necessary for single-cell visualization.

The development of *fluorescence* technology in biological sample preparation has revolutionized cell biology and, most recently, DNA manipulation techniques such as the DNA chip from Affymetrix, and has given rise to optical brain monitoring techniques. The interdisciplinary approach — combining chemistry and biology — has energized both scientific disciplines. *Material sciences* are combining chemistry and biology in ways different from those in classical biochemistry. Nanotechnology, the manipulation of matter for a single molecule, is a new development and has become a buzz-word, indicating that technology is improving the miniaturization of mechanical and electrical devices, with the goal being single-molecule machines. The nanometer, the physical dimension of this new world, is the dimension of the molecular machinery of life itself. Hence, inspiration and creative solutions among biologists, chemists, and engineers must go hand in hand.

Medicine is another beneficiary of computer technology. Non-invasive techniques eliminate the need for surgical procedures, the use of chemicals, or the use of X-rays. Two successful non-invasive techniques are functional magnetic resonance imaging (fMRI) and ultrasound which make use of the physical properties of specific atoms or molecules shared by all matter. In addition, the development and deployment of expert systems will increase the ease and success of performing routine medical treatments.

Computers in Neurobiology and the Mapping of the Human Brain

The brain receives attention from scientists for obvious reasons: behavior, consciousness, memory, sleep disorders, perception, or pain, including that in phantom limbs. Recent computer-aided brain scanning techniques and the success of the Human Genome Project have revitalized attempts to map the human brain both anatomically and functionally. With these maps, it should one day be feasible to co-localize neuronal activity at the level of single sensory neurons as a function of motoneural tasks, learning words, or reading symbols. The newly coined term *neuroinformatics* promises for neuroscience what is currently happening for the human genome, namely, the integration of maps (or sequences) with functional data. Such integration produces *mechanistic* explanations for human behavior. The availability of the information through computer networks like the Internet is a vital component of this undertaking (see *Science* online, 279:1320 at http://www.scienceonline.org/). On April 2, 1993, the Human Brain Project was officially announced by the NIH. The Human Brain Project[1] is designed as a broad-based, long-term research initiative to support research and development of advanced technologies and to open information superhighways (e.g., Internet) to neuroscientists and behavioral scientists (the Human Brain Project, http://www-HBP.scripps.edu/).

The Human Brain Project consists conceptually of three sub-projects: the Multi-Modal Imaging and Analysis of Neuronal Connectivity or *Connectivity Project*, the *In Vivo* Atlases of Brain Development or *Atlas Project*, and the Goal-Based Algorithms for 3-D Analysis and Visualization or *Algorithms Project*. A number of distinct technologies need to be integrated into these projects, such as chemistry, animal models, computation, Web design, networking, and functional magnetic resonance imaging. These technologies include the design, synthesis, purification, and characterization of novel contrast agents (chemistry core); animal acquisition, care, breeding, surgery, and disposal (animal core); design and maintenance of a stable and effective computational environment; implementation of algorithms associated with goal-directed imaging experiments; dissemination of those results to provide routine hardware and software support for computers, networks, Web page maintenance, and video environments; and finally, the development and maintenance of an imaging facility that uses nuclear magnetic resonance (NMR). To better understand the complexity of the human brain, the ultimate goal of the Brain Project is the "weaving of the informatic and neuroscience components" of brain research into one single unit. According to the California Institute of Technology:

> ...past progress in this Human Brain Project has been fueled by ongoing and dynamic interactions across the various individual Projects and Cores. This is aided by the fact that many of the personnel have expertise and interests that cross the artificial Project/Core boundaries. Work in the informatics components provides the neuroscience components with novel and more efficient ways of collecting, analyzing, and looking at information. This necessitates the computational components learning from the neuroscience components about how the data is collected, as well as what it is about the data that is interesting and important. Thus, we have a win-win situation where each component aids and is in return enriched by the other component. (source: http://www.gg.caltech.edu/hbp/)

Neurobiology is becoming the playground for a wide range of specialists, from molecular biologists and electrophysiologists to cognitive scientists and philosophers. There are those who pursue epistemological questions related to consciousness, believing that science on the molecular level will ultimately reveal the "mystery" of the mind–body problem. Brain research is divided among myriad specialists, each having a singular approach, technique, and methodology. There is no common language that stretches from biophysics to psychology, although many scientists believe that such a common language — a unified field — will one day exist. The complexity of the brain and the effort to understand it are enormous. Feelings and analytical thinking, two central aspects of human consciousness, are elusive properties from a biological point of view. It is true that certain chemicals influence our consciousness showing that it is firmly rooted in the central nervous system. However, very little is known about potential mechanisms that reveal how consciousness emerges from brain chemistry (spatial and temporal). Neurobiology suffers from large gaps in information that still exist among some of the life science disciplines. Saying that the whole is more than the sum of its parts is true, yet no model exists that could explain exactly how such a transcendence of these functional levels might occur.

In their need to understand the function of the brain or of a protein folding into an active enzyme after its synthesis, scientists want to adopt a new perspective — that of a water molecule interacting with the folding protein. Here again, computers can help generate functional *landscapes* in which the desired parameters can be visualized (e.g., the electrostatic forces between atoms and molecules), a view utterly foreign to our experience and thinking. We are aware only of our own experience, and what we lack in our attempt to understand a water molecule or virus is the very consciousness of being one. A similar argument about the "radar" capability of bats was made by the philosopher Thomas Nagel. For example, we have learned to avoid putting our fingers in sockets for fear of being electrocuted. Although we experience the effects of gravity, we have no sense that allows us to detect an electromagnetic force.

To overcome our inability to "see" the atomic world and force fields outside of man's experience, we must create images of our own — translations, so to speak — to recreate what we think happens. *False color imaging* is one such powerful tool. The units of a physical parameter are transformed into a color code from red to blue. Because we do not have to read and compare numbers, our minds instantly "see" differences and identities distributed in space. This way, we can "see" temperature gradients (visualized infrared imaging) or molecular oxygen consumption in the brain to localize spots of high metabolic activity. False color imaging is a fascinating and aesthetically satisfying technique because it allows us to instantly transfer abstract mathematical formalism describing *numerical* relations — and science is all about numerical values — between objects into direct sensory inputs for our visual receptors (eyes). Here again, computational power is instrumental in the implementation of these graphical tools.

An integrated map of the human brain superimposes an anatomical and functional map both in space and time. The types of questions that can be asked depend on the level of resolution. For improved visualization of the spatial arrangements of functional brain units, topological maps, like cartography of the landscape, aim to produce rows of cross sections layered on top of each other. Functional brain mapping makes use of non-invasive techniques such as SPECT/PET (single photon emission computed tomography/positron emission tomography), fMRI (functional magnetic resonance imaging), EEG (electroencephalography), MEG (magnetoencephalography), optical imaging, and neuroanatomical tools. These tools are then used to produce maps composed of cross-sectional images of the brain. Cross sections are put back together into a virtual 3-D image reconstruction of the brain.

Spatial resolution in the millimeter range (fMRI) and temporal resolution in the millisecond range (EEG, MEG) need to be correlated by computational means and result in movies of brain activity (Liu et al., 1998, PNAS 95:8945 at http://www.pnas.org/). These brain maps have a resolution at the *neuroanatomical* level. *The Whole Brain Atlas* at Harvard Medical School (http://www.med.harvard.edu/AANLIB/home.html) is an example of a publicly available real-time map that compares normal brains with those affected by cerebrovascular diseases (strokes), neoplastic diseases (tumors), degenerative diseases (Alzheimer's, Huntington's), and inflammatory or infectious diseases (multiple sclerosis, AIDS-related dementia, Creutzfeld-Jakob, herpes). These maps reflect the high spatial resolution obtained from hemodynamic or metabolic measurements, such as glucose levels and oxygen consumption with the high temporal resolution of their electromagnetic signals. This correlation is not trivial because the brain is a complex organ with anatomically distinct processing centers that communicate across large distances and several time scales (Liu et al. 1998).

However, spatial resolution in millimeters is by no means good enough for molecular studies of brain functions. Although neurons outgrow axons measuring several millimeters to meters (peripheral nervous system), the thickness of their cell bodies must be measured in micrometers or less when studying the functional morphology of their chemical synapses. Thus, neuroanatomical maps like *The Whole Brain Atlas* can be complemented with molecular details stemming from biochemical, physiological, pharmacological, and molecular biological studies, such as ion channel and receptor distribution (proteomics), mRNA level distribution (genomics), synaptic connectivity and plasticity responsible for the enormous complexity of neuronal networks. While the current information content and resolution level of *The Whole Brain Atlas* can pinpoint anatomical centers of activity for motoneuronal tasks, for example, they are not yet able to provide information about electrical activity patterns (firing patterns) and neurotransmitter selectivity of individual neurons or groups of neurons. In other words, the structural details are not yet linked to functional states, the temporal activity of the brain.

Computer-generated virtual images allow the view of objects from a perspective that we are otherwise unable to have. We can now think of ourselves as agents small enough to roam the inside of a body (movies have been made attempting just that, promoting an artistic view of the inside of man). In order to illustrate the relationship between structure and function, imagine a new house you are planning to build with an architect. You can see yourself inside using the space and going from room to room. It is not necessary to create a cross-section map of the house for visualization purposes because we know how its structure and function are related. We are familiar with the sensation of walking into the kitchen and having the aroma of dinner awaken our senses, knowing that we are about to eat our favorite foods. We often like a particular food, not only for its taste, but because of the person who cooked it.

We do not have this insight for objects in biology. Understanding a fundamental experience like enjoying eating differs dramatically from scientists trying to understand a virus. As much as we can understand the joy of eating experienced by others, we obviously cannot use an analogy to understand a viral infection from the point of view of the virus. We are left with the mere mathematical description of a process of molecules interacting, virus particles "docking" to cell surfaces. And as much as we are aware of a food when eating it, we are totally oblivious to the process of digestion and nutrient absorption (unless there are problems). Our awareness of metabolic activity is as nonexistent as our ability to feel like a virus. Most of our bodily functions are beyond awareness, and it is a good guess that microbial life happens at a similar unconscious level. We do not feel or think with our liver, but with our brain. If only we could understand the difference.

From Molecules to Neurodegenerative Diseases

Although we have conscious experiences, we do not have a mechanistic model for them. Neurobiology, however, has made great progress in characterizing the basic function of individual neurons and their components. The propagation of action potentials in neurons is a well-understood, basic signaling mechanism of the brain. Action potentials are localized, small voltage changes (measured in millivolts) across the cell membrane and their lifetime is measured in milliseconds. Repetitive action potentials form *firing patterns* in individual nerve cells. These firing patterns are the intrinsic signaling mechanisms of neurons and differ for different types of cells. How are they measured? Because of their short lifetime, a handheld stopwatch is an inadequate instrument for measurement. While analog instruments for recording electric activity of cell membranes are fast enough to record such activities on high frequency chart recorders, scientists are left with the arduous task of analyzing the data by hand — that is, taking a ruler, measuring time and signal amplitudes on the chart recording paper, and then entering the numbers into a calculator for statistical analysis. Computer-aided simulation, real-time recording, and the immediate input into analysis software has shortened the time needed for measuring and interpreting data by several orders of magnitude. Data analysis can be performed in minutes instead of hours or days by using commercially available spreadsheet programs.

Molecular neurobiology addresses the structure–function relationship of so-called ion channels, the protein complexes responsible for action potentials. Many of these channel proteins can be linked to neurodegenerative diseases and susceptibility to drugs and toxins. Although ion channels are well studied both functionally and structurally, with the first studies done in the 1940s, the use of molecular biology to unravel novel genes and genomic organization of channel proteins during the 1990s has greatly improved our understanding of Alzheimer's, Parkinson's, and Huntington's diseases. The Human Genome Project will further our understanding of those and other hereditary diseases. The search for treatments and cures is a fertile testing ground for bioinformatics tools at almost every level of investigation.

Ion channels provide narrow, channel-like pathways used by ions to cross cell membranes. The membrane is an electrical insulator and blocks the movement of ions (positively or negatively charged molecules) in the absence of channels. Ion flux can be measured as an electrical current. With extremely sensitive electronic amplifiers, the movement of a few thousand ions in every thousandth of a second can be measured and could reveal information about the electrical activity of a cell membrane or within a narrow patch thereof. Action potentials, the signaling mode of neurons, are the combination of at least three different classes of ion channels, meaning that for each of the different ions, a different protein has to be present for it to cross the membrane. Ion channels are known to be ion selective.

Because of the nature of electrical activity and the extremely low number of charged molecules moving across membranes, computers have become indispensable tools in the control and analysis of electrophysiological experiments. It was the development in 1975 of micro glass pipette techniques to measure the activity of single proteins that revolutionized neurobiology and pharmacology. Electrophysiology was born in the 1940s and has grown into an industry where scientists interested in particular proteins or diseases can study specific pharmacology and physiology through commercially available computers that help push the limits of experimental systems, especially in terms of very rapid events. Current signals that can be converted into voltage signals or coupled to fast optical events, allowing ever-improving time and amplitude resolution, are reaching the theoretical limit of measuring the activity of a single molecule.

The 1940s was an era of analog machines. Experience and intuition had already changed the understanding of neuronal activity from macroscopic to microscopic. Two decades earlier, physics and chemistry had unraveled the mystery of the atomic structure and the nature of the chemical bond. The investigation of the structure and function of single molecules in biology thus seemed only a matter of years away. Electron microscopy was already showing the shape of single cells and viruses. In neurobiology, much of what an ion channel's activity is all about has been inferred from biochemical and kinetic analysis of electric currents of single cells. The concept of unit structures from particle physics certainly helped in shaping the idea of the existence of single channels formed by proteins, of which several make up the observed macroscopic behavior of nerve cells. This requires correct mathematical formalism and little has changed, except for the speed and number of parameters that can be handled experimentally. Of course, this makes a huge difference in understanding complex systems.

A scientist no longer needs to be a pioneer in electronics and code writing to utilize computers. Scientists finally have a tool to study the function of proteins responsible for brain activity at the molecular level. Most known drugs act at the level of these ion channel proteins or associated proteins that control the activity of these channels. Also, most potent, and often lethal, toxins from snakes and spider venom directly bind and interfere with the functional activities of these channels.

Long-QT syndrome is a defect in the rhythm control of the heart muscle as the result of structural abnormalities in the potassium channels of the heart, which predisposes affected persons to an accelerated heart rhythm.[2] This can lead to sudden loss of consciousness and may cause sudden cardiac death in teenagers and young adults who are faced with stresses ranging from exercise to loud sounds (source NCBI; genes and disease page).

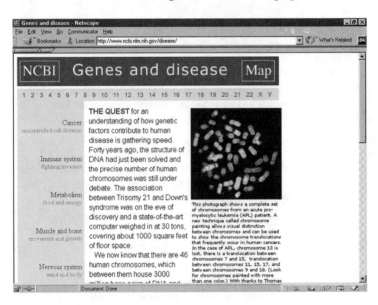

FIGURE 5.2
Genes and disease: a database at NCBI for genes involved in hereditary diseases

The responsible protein is a potassium-selective ion channel which excludes all other biologically important ions like sodium, chloride, calcium, and magnesium. Potassium is found mostly inside cells and outside only at a tenfold-lower concentration. The effect is that whenever a potassium channel opens, potassium ions move from the inside of the cell to the outside, moving positive charges to the extracellular space. This movement will happen as long as the channel is open and until the potassium concentration is balanced — that is, equally concentrated inside and outside. Yet this equilibrium never occurs in normal cells.

There are two mechanism that could explain why this does not happen in healthy cells. First, they are able to pump the potassium ions back into the cell while simultaneously shuffling sodium ions out of the cell through the same transport proteins, creating an asymmetric distribution of sodium ions which generates an electrogenic force opposing the flux direction of potassium. Second, after being open a very short time, the potassium channels are regulated and permanently closed (inactivate). The time and amount of current flowing through potassium channels varies from cell to cell, which express closely related, but distinctly different potassium channels. This is possible because different potassium channel genes, although closely related, coexist in a cell's genome. These genes are differentially expressed, which results in cell or tissue-specific potassium channel distribution and activity. The net effect of potassium channel activity is the repolarization (restoration) of the resting potential of neurons. In the Long-QT-syndrome, this process is markedly slowed because of mutations in one of the potassium channel genes causing induction of arrhythmia in the heart muscle.

So far, many potassium channel genes have been found by using a clone-by-clone approach. Sequence analysis and comparison of these genes allows us to better understand their role and how organisms make use of these proteins. The combination of DNA sequence analysis (and for that matter, amino acid sequence analysis) with functional information is tremendously important in furthering biological knowledge. Associating function with genetic variability and cell type-specific expression and use will lead to the building of functional map databases.

How many potassium channel genes are there? The current number of cloned and sequenced genes is more than 50, a number that includes genes from different organisms — from man to bacteria. The growing numbers of sequences allow interesting interpretations that are based simply on their sequence differences with well-elaborated information about their functions. Once again, the relationship between sequence, structure, and function is enormously helpful in our understanding of the physiology, including diseases, of these channels.

Physiology, pharmacology, molecular genetics, and phylogenetic analysis (comparing channels not only within the same organism, but between such diverse species as man and plant) are the future of molecular biology, where the pieces of the reductionist puzzle are being put in place, one by one. This puzzle will eventually lead to the understanding of how drugs work, not only in an *in vitro* assay where binding studies demonstrate specificity and affinity, but the understanding of how a drug that interacts with one kind of potassium channel exerts a mind-altering or pain-suppressing effect, because a phenomenon like pain is not understood at the molecular level. Finding drugs that act as painkillers is like searching for a needle in a dark cellar using the light from a fluorescent wristwatch. Two categories of known diseases linked to ion channel mutations (defects) are listed in Table 5.1. Ion channels are also susceptible to toxins and drugs, such as calcium channels for snail toxins and caffeine, acetylcholine receptors for nicotine, and potassium channels for the snake toxin charybdotoxin.

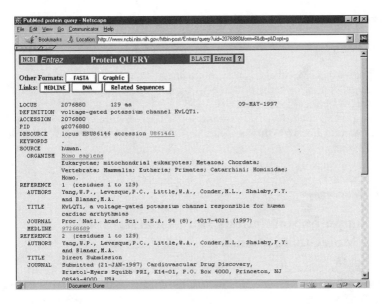

FIGURE 5.3
PubMed protein query for voltage-gated potassium channel KvLQT1: this protein sequence contains a mutation responsible for the disease known as Long-QT syndrome which produces human cardiac arrhythmias.

TABLE 5.1

Known Diseases Linked to Ion Channel Mutations

Diseases	Susceptibility to Toxins and Drugs
Gap junction protein Cx43; X-linked Charcot-Marie-Tooth disease	Calcium channels; omega-conotoxin (snail toxins), caffeine
Chloride channel ClC-2; Myotonia	Acetylcholine receptor; nicotine, alpha-bungarotoxin
Long-QT syndrome	Potassium channels; charybdotoxin

For more information, consult the National Center for Biotechnology Information page on genes and diseases at http://www.ncbi.nlm.nih.gov/disease/

Sequencing and gene mapping progress resulted in the identification of the genes responsible for the diseases listed in Table 5.1. The National Center for Biotechnology Information offers a summary link to genetically determined diseases (http://www.ncbi.nlm.nih.gov/disease/). Eight distinct inherited diseases are linked to ion channels, pumps, and transporters, including cystic fibrosis (chloride channel), diastrophic dysplasia (sulfate transporter), Long-QT syndrome (potassium channel), Menker's syndrome (copper transport), Pendred syndrome (thyroid-specific sulfate transporter), polycystic kidney disease (cell–cell interaction, membrane protein organization), Wilson's disease (copper transport, ATPase), and Zellweger syndrome (PXR1, peroxisome protein import receptor, peroxisome biogenesis disorder).

References

1. Shepherd, G.M., et al., The Human Brain Project: neuroinformatics tools for integrating, searching and modeling multidisciplinary neuroscience data. *Trends Neurosci.*, 1998. 21(11): p. 460-8.
2. Yang, W.P., et al., KvLQT1, a voltage-gated potassium channel responsible for human cardiac arrhythmias. *Proc. Natl. Acad. Sci. U.S.A.*, 1997. 94(8): p. 4017-21.

5.2   Computer Simulations and Visualization of Molecular Structures

Biology has become an air-conditioned enterprise. Laboratories in research universities, private research institutions, and the pharmaceutical and bio-technology industries are filled with expensive equipment and busy people handling small quantities of solutions that are mostly translucent, yet contain those essential molecules known as DNA (deoxyribonucleic acid) and protein. Modern biology has become a science of handling molecules of life that are invisible to the naked eye. Unlike Gregor Mendel, who described the colors and shapes of peas growing in his garden, today's scientist using a high-powered microscope can visualize the location of the actual genes causing a pea to become green or yellow, or plump or shriveled, and can determine their molecular structures by using indirect detection methods. How do we know that genes look like the famous double helix described by Watson and Crick in 1953? What are the structural features of known proteins in today's biological world? Since a molecule cannot be seen by microscopic means, its shape has to be inferred from experimental data through mathematical analysis.

3-D Visualization

The chemical structure of all matter was first studied and solved using high energy radiation (X-rays) that caused diffraction into regular patterns when beamed onto a crystalline aggregate of this molecule. The diffraction pattern and intensity of the diffracted X-ray beams can be used to calculate the actual distribution of electrons in the molecular crystal. The complex mathematical formalism makes it impossible to do calculations by hand. The use of computers, therefore, has been instrumental in elucidating the structures of complex molecules such as proteins, nucleic acids, lipids, and carbohydrates. The sheer number of atoms involved in biological systems limits man's ability to calculate, and due to the limitations of computational power and speed, many modern problems cannot be solved even with the newest supercomputers.

Real molecules can be represented by macroscopic values. Viscosity of water, for example, represents the dynamics of intermolecular hydrogen bonding between water molecules in the fluid phase and the size of a moving particle. Thus, the size and molecular weight of a protein can be determined by measuring the viscosity of a protein solution. The biologically significant interactions (covalent and non-covalent bonds) between molecules are tremendously complex if they are to be described in quantum mechanical terms, but the dielectric constant, a single physical parameter measuring the electrical polarization of water, simplifies the calculation by treating the interaction of charged molecules in water. For a biochemist, the dielectric constant of water can be used to describe the energy of interaction of a protein surface with water or a protein or DNA surface. Similarly, membrane proteins can be dealt with using a single parameter reflecting the dielectric properties of hydrophobic solvents, which differ dramatically from aqueous solutions. This limitation — which, in effect, is a practical simplification — is why, in computer simulations of protein structures and dynamics, the solvents (e.g., water molecules) are replaced for a single macroscopic value. Otherwise, searching for a solution in a reasonable time, say a few weeks, would be prohibitive due to the large number of independent parameters.

As an example, Figure 5.4 shows the structure of the human immunodeficiency virus (HIV) reverse transcriptase. The structure of this enzyme has been solved to high resolution (2.6Å) describing the spatial coordinates of 7715 atoms, not including all hydrogen atoms. A total of 27108 reflections have been measured for structure refinement using the program X-PLOR 3.1.[1] Using the accession number 1REV at the query page of the Brookhaven Protein Data Bank[2] (PDB; http://www.pdb.bnl.gov/pdb-bin/pdbmain), a summary page indicates date of submission (9/17/1995), title of protein (HIV reverse transcriptase), and the authors (J. Ren et al.).

Classification information about compound and organism gives the location from which the protein originates (it is an engineered, or recombinant, enzyme with the enzyme commission number EC 2.7.7.49, and is an enzyme of the human immunodeficiency virus type 1), and has been expressed in a bacterial system to maximize quantities for crystallization purposes (the expression system of choice is an *E. coli* cell into which the recombinant DNA containing a gene of a virus that can only infect humans and not bacteria, has been inserted in order to overexpress the protein for purification). Protein crystal parameters (space group), methods, and additional ligands and resolution (2.6 angstroms) are also indicated. To get a complete understanding of the current structure of this protein, there are several links that will enable the scientist to access the complete coordinates in spreadsheet format and 3-D and 2-D visualization. When following the link (header only) under "data retrieval," additional information and links (to the published article in MEDLINE) and sequence information in GenBank[3] and SWISS-PROT[4] can be found.

"View 1REV in 3-D" gives a link to MDL's Chemscape Chime viewer (http://www.mdli.com/chemscape/chime/), a software plug-in for easy viewing

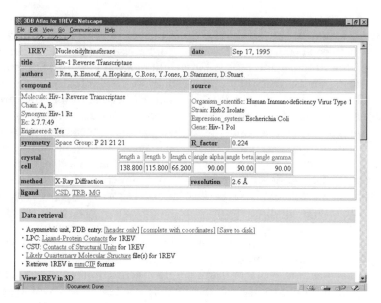

FIGURE 5.4
HIV reverse transcriptase data sheet at PDB

of the structure in different modes. Alternatively, the PDB structure file can be downloaded onto the local hard drive and viewed by the Rasmol browser plug-in Chime (Figure 5.5).

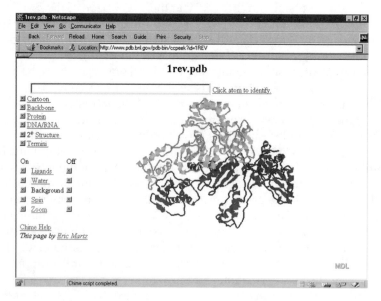

FIGURE 5.5
HIV reverse transcriptase structure 1REV.PDB viewed with Chime

It is interesting to look at the history of generating images of biological macromolecules. An historical outline by Eric Martz can be found online at www.umass.edu/microbio/rasmol/history.htm. Computer representation of wireframe models was not possible until 1972 and was pioneered by Levinthal and Katz. Years later, computer generation of models of spacefilling molecules and the rotation of the molecules become possible. In recent years, inexpensive programs have become available for use in classroom settings which can be operated on Macs or PCs with less than 1Mb of disk space (however, the data files use hundreds of kilobytes and could easily use up considerable disk space if a personal structural database were built on a home computer). Yet programs like Kinemage[5] and Rasmol[6] demonstrate the ease of looking at protein structures (Rasmol is freeware).

The Virtual Doctor (Virtual and Remote Surgery)

Computers and robots are becoming integral partners in advanced medical procedures. The time efficiency and increased accuracy of these procedures are driving their advancement in today's as well as tomorrow's medicine. Emulating software is now used in surgery rehearsals and training at certain university hospitals and is a training tool for many surgical residents in different specialties and subspecialties. Virtual surgeries could improve the success of certain procedures, but are limited by their inability to anticipate problems that could arise during the actual surgery. This, in turn, will better prepare the surgical staff and will improve the surgery's outcome. Therefore, it is crucial for the emulating software to be as realistic as possible. To ensure realistic volume visualization of the regions displayed, the specification and representations of the anatomical regions in 3-D space must be very accurate. During the virtual surgery simulation, the program must be able to detect and quantify changes at the operating level in order to manipulate the data output that enables the operator to visualize the anatomical region and the volume of interest.

Visualization of the human anatomy has been a great challenge for many centuries. Leonardo da Vinci's detailed anatomical drawings (dating from 1489) had important educational, diagnostic, and therapeutic impact during his lifetime. What Leonardo's paintings lacked was *in vivo* representation, which was not made possible until Wilhelm Conrad Roentgen discovered X-rays in 1895. X-rays were used to create the first *in vivo* medical imaging technique. This method was limited to 2-D representation of the cross sections analyzed, and the information it gathered yielded little information about the organs themselves. Nevertheless, X-rays are very powerful medical tools that are still used today. The development of computerized tomography (CT) in the 1970s and magnetic resonance imaging (MRI) in the 1980s has enhanced cross-sectional imaging of anatomical features. The 2-D images obtained from these techniques have been instrumental in diagnosis and have also helped to improve anatomical teaching tools.

The drawbacks of 2-D images are their lack of volume information and the relationship of the features displayed in 3-D space. 3-D images would add a great deal of realism and relativistic information (tissue–tissue or organ–organ interactions). In a 2-D image, the information obtained is based on a two-coordinate system (x,y). In order to generate 3-D images from these 2-D pictures, information from a third dimension (z) is necessary. Therefore, multiple 2-D images with fixed x,y coordinates and a variable z coordinate are used to generate these three-dimensional images. Several 3-D reconstruction techniques have been employed to improve the quality and realism of the images portrayed and to allow the observer to get further involved in the displayed image. The "gray level gradient shading",[7] the "generalized voxel model",[8] and the reconstruction of the "visible human" model are a few of the most successful projects of the last decade. With the gray-level gradient technique, the computer uses the original tomographic data to compute a smooth surface normal with a high dynamic range. The generalized voxel model allows further exploration of the imaged data displayed. An example of its application is showing different organs in combination with others and allowing selective cutting of the 3-D images displayed. The most recent efforts have been concentrated on constructing the virtual "visible human".[9] In this project, the use of colored, cross-sectional images has enhanced the 3-D quality of the displayed object and has added a great deal of realism to the features portrayed. Computers have greatly enhanced medicine and they are here to stay.

References

1. Badger, J., et al., New features and enhancements in the X-PLOR computer program. *Proteins*, 1999. 35(1): p. 25-33.
2. Sussman, J.L., et al., Protein Data Bank (PDB): database of three-dimensional structural information of biological macromolecules. *Acta. Crystallogr. D. Biol. Crystallogr.*, 1998. 54(1 (Pt 6)): p. 1078-84.
3. Benson, D.A., et al., GenBank. *Nucleic Acids Res.*, 1999. 27(1): p. 12-7.
4. Bairoch, A. and R. Apweiler, The SWISS-PROT protein sequence data bank and its supplement TrEMBL in 1999. *Nucleic Acids Res.*, 1999. 27(1): p. 49-54.
5. Richardson, D.C. and J.S. Richardson, The kinemage: a tool for scientific communication. *Protein Sci.*, 1992. 1(1): p. 3-9.
6. Sayle, R.A. and E.J. Milner-White, RASMOL: biomolecular graphics for all. *Trends Biochem. Sci.*, 1995. 20(9): p. 374.
7. Cao, Q.L., et al., Enhanced comprehension of dynamic cardiovascular anatomy by three-dimensional echocardiography with the use of mixed shading techniques. *Echocardiography*, 1994. 11(6): p. 627-33.
8. Schmidt, R., et al., Visualization of 3-D treatment plans with fast neutrons. *Strahlenther Onkol.*, 1992. 168(12): p. 698-702.
9. Slavin, K.V., The visible human project. *Surg. Neurol.*, 1997. 48(6): p. 638-9.

5.3   Predictive Biology

Our most recent advancements in biology and medicine are enhancing the predictive potentials of these fields. The possibilities introduced with such an understanding of the life sciences are endless. Our enhanced ability to predict outcomes in biological experiments has improved and will continue to improve our living standards, thus allowing many science-fiction topics to become reality. This occurs mainly through advancements in therapeutics (e.g., pharmaceuticals), and these enhancements in drug design will undoubtedly improve and prolong life. Although the majority of these discoveries are designed to enrich and enhance our daily activities, the accompanying negative and potentially dangerous implications are inevitable.

One of these controversial issues is that of cloning human beings. Each human being is unique in the sense of the genetic information he or she carries — identical twins being an obvious exception. Nevertheless, it is this uniqueness that sometimes hinders our survival (i.e., preventing us from accepting organs from fellow donors). One rationale for human cloning is that it could alleviate many of the problems associated with our unique characteristics. The genetic information (DNA) of a cloned organ is identical to that of the individual needing the organ. Hence, the cloned organ would be accepted by the individual's body as part of itself and would have a smaller probability of rejection. In order to achieve such goals, we must be able to overcome certain obstacles. Analyzing the entire genome of the individual is a necessary step in this process, and would be impossible without powerful computational tools. Computers and robots will be able to analyze the relevant data and perform the tasks necessary for achieving such incredible goals.

It was not that long ago that pharmaceutical science could only follow an empirical approach for drug design. Thousands of different compounds are tested every day to find suitable drugs for diseases of interest. The lack of applicable biological laws forces such trial-and-error approaches. Today, the tremendous strides in the biological sciences have set forth biological and biochemical laws that are specifically applicable to the science of drug design. Prior to the experimental testing phase, a compound library with thousands of potential compounds can now be narrowed down by several folds. This, in turn, has enabled a more efficient high-throughput screening to search, in large molecular libraries generated by combinatorial chemistry, for candidates that could potentially serve as life-saving therapeutic agents.

Over the past several decades, the exponential growth in biological and biochemical data at the molecular level has been one of the greatest contributors to the transformation of pharmaceuticals from an empirical state to an approach of rational and systematic design. A necessary ingredient in rational drug design is knowledge of the molecular structures involved — both the receptor protein and the drug molecule. In the absence of known structures — and this is true for many of the novel genes discovered through the genome projects as open or unidentified reading frames — structure prediction on the side of the receptor is the only viable alternative for such a rational approach.

Protein Structure Prediction

Modeling biological macromolecules was initiated in the 1940s, long before advanced computational tools were available. Linus Pauling, James Watson, and Sir Francis Crick were the key pioneers (in a field of many excellent biochemists and crystallographers) who in 1953 contributed to the successful solving of the structure of DNA, and the first protein structure in 1961. Modeling the structure of biological macromolecules allows us to gain a great deal of insight into the molecule's functional features.

The first few biological macromolecules modeled were the double-stranded DNA helix (a polynucleotide) and the oxygen-binding protein myoglobin, an alpha helix containing protein. The absence of advanced computational tools made these modeling efforts very time consuming. Nearly ten years of experimentation and extensive computation were required for the modeling efforts for the alpha helix and the double-stranded DNA.

Linus Pauling was the father of molecular modeling and his contributions to this field were instrumental to the work of the scientists who followed in his footsteps. His modeled structure of the alpha helix enabled the identification of structural patterns, called secondary structures, that exist at the molecular level of proteins in all organisms as shown in Color Figure 1.* This, in turn, helped to recognize structural consensus (secondary structures) in other biological macromolecules and enabled initiation of a classification scheme associated with the polypeptide's secondary structures (alpha helix, beta strand; currently over 250 types have been classified, Holm et al., Mapping the protein universe, 1996, *Science* 273:595-602).

Sir Francis Crick and James Watson were awarded the Nobel prize for solving the structure of the double-stranded DNA molecule. The structure of this molecule was an essential step in solving many of the ambiguities associated with biological inheritance and development and enabled the development of a new branch of biology that dealt specifically with biological problems at the molecular level: molecular biology. This field is mainly concerned with understanding the DNA molecule in a variety of species and its direct and indirect interactions with other biological macromolecules (e.g., proteins). The vast amounts of information associated with the structure of the DNA molecule enabled us to gain incredible insight into the phenotypic (physical) and molecular characteristics of many different species.

---

* Color Figure 1 follows page 52.

In the early part of the twentieth century, the notion of a molecular understanding of biological systems was considered to be farfetched and something that could be found only in science-fiction books. A few decades later, Avery's experiments dismissed the notion of proteins as the coding molecules (blueprints) in living systems and supported the idea of deoxyribonucleic acid (DNA) as the responsible coding agent. This evidence created an urgent need to understand every intricate detail of this molecule. This, in turn, attracted many brilliant scientists who continue to contribute to the field's exponential growth and advancements.

Today, X-ray crystallography, nuclear magnetic resonance (NMR), and cryo-electronmicroscopy enabling electron diffraction are the techniques used in solving the high-resolution stuctures of biological macromolecules. It was X-ray crystallography that first determined the structure of the DNA molecule. This technique was also utilized to verify the modeled structure of Linus Pauling's alpha helix in proteins.

In X-ray crystallography, the crystallized protein is bombarded with electrons and its electron diffraction pattern is used to determine the atomic structure of the molecule. Molecules are composed of atoms, which are further composed of electrons, protons, and neutrons. Different atoms have distinct electronic signatures that enable us to distinguish one from another. The electron diffraction pattern is then used to calculate the atomic coordinates based on the measured electron density associated with each of the detected atoms. In principal, since this method is not limited to the size of the molecules, it could be used to determine the structure of any macromolecular structure — even of entire cells if they were to aggregate within a crystal in an orderly fashion of millions of units.

There are, however, many limitations associated with its methodology. First of all, crystallizing the biological macromolecules is not a trivial step, mainly due to the difficulty of obtaining regular crystal lattices of millions of identical units. Many of our essential proteins (e.g., membrane-bound proteins) are still structurally unknown due to the inability to crystallize these polypeptides once they are removed from their membrane environment. Electron diffraction, however, is an excellent alternative thanks to cryotechniques, where samples are frozen at extremely fast rates (milliseconds) in liquid nitrogen (–180 degrees C).

A second problem includes the extensive calculations associated with the analysis of our data set. This problem was alleviated by the introduction of advanced computers. The lack of dynamic information is another limitation in obtaining structures through crystallography. Biological macromolecules (e.g., proteins) are mainly found in aqueous environments and their structures are quite dynamic and flexible. Crystallography gives us a rigid picture of the molecule due to the very long time exposure needed to collect the diffraction patterns.

NMR data, on the other hand, yields all possible modes of conformation, since its much higher time resolution is in the millisecond range. Although the use of NMR seems to be advantageous and the most reasonable technique for solving structures, its limitations necessitate alternate techniques (e.g., X-ray crystallography, electron diffraction). The use of proton signatures in NMR creates noise (overlapping signatures) in larger molecules. This limits it to smaller macromolecules and prevents the analysis of molecules greater than 30 KD. Analyses are now performed with the aid of computer programs. The extensive, time-consuming calculation steps involved in crystallographic (e.g., Fourier series, etc.) and NMR analyses are now performed with incredible accuracy by automated and semi-automated programs in a fraction of the time it took to perform them manually.

Today, the rate at which sequence information is published each day far exceeds that of structural information obtained through crystallography and NMR. Hence, the need for and popularity of predictive structural biology has been enhanced dramatically and is utilized often in a variety of scientific applications (e.g., pharmaceuticals, etc.). In the past, this sort of biology was practiced predominantly by theoreticians in the biochemical, biophysical, and biomedical sciences. These investigators later incorporated advanced computational tools to investigate biological problems that might have been explained through physical and chemical laws. The computers allowed them to automate some of these tasks and enabled them to identify some fascinating physical and chemical trends present in the biological systems studied. This was a major step in the advancement of predictive biology. Without computers, the comparative step of this process would be dramatically hindered. With today's computers, the analysis of a typical protein molecule can take mere days — the absence of such tools would add years to this step. The need for computational tools and the advancements achieved through their application has given rise to new, emerging fields, including computational biology and computer science in medicine.

Understanding the structural aspects of the protein of interest will yield a vast amount of information about its potential function and its relationship to other essential macromolecules. Predicting the high-resolution structure of proteins and their correct folding route, however, is a major problem in biology, because proteins are composed of twenty different alpha amino acids that introduce the essential sequence variability observed in these molecules. Each of these amino acids in turn can hold a variety of conformations with respect to the other residues (the side chain associated with each of the amino acids) in the protein.

The number of different structural possibilities to which the protein can conform with respect to its amino acid sequence are phenomenal in theory, but only one or several closely related active structures are formed in solution. Many different methodologies have been employed to explain why only one preferred conformation is found from these astronomical numbers of conformational possibilities. The most common deductive techniques used are those dealing with energetics (thermodynamics) and those that take advantage of the molecule's relational information to other known homologs. Currently, the consensus in the field is to employ both methods in an ordered fashion. When possible, homologous molecules would be used to gain an insight into the molecule of interest and the use of energy minimization software would be employed to minimize chemical and physical anomalies of interaction as shown in Color Figure 2.*

As discussed earlier, the rate of solving structures through X-ray crystallography or NMR is much lower than the high numbers of new DNA and protein sequences that are introduced each day, necessitating a calculated and semi-reliable approach to structural design. The bioinformatics groups are predominantly responsible for identifying key biological macromolecules (e.g., proteins) responsible for pathological events and for proposing potential inhibitors for such macromolecules. In order to understand and, thus propose the possible interactions of the molecule of interest with its substrate and other interacting compounds, the structure of the macromolecule studied must be known. This introduces several problems. The greatest obstacle is the lack of structural data for the molecule of interest. In most cases, the macromolecules studied are structural or regulatory proteins. For proteins that lack NMR or crystallographic structural information, homologous proteins from other species could be used for which a structure is known. Modeling unknown protein structures based on their homologs is better known as homology-based structural modeling and the program that best exemplifies such an approach is the homology program from MSI's InsightII software programs. In the next several pages we will discuss the details involved in the homology-based modeling approach and point out its most obvious strengths and limitations.

Many of the essential proteins present in humans are also found in other living organisms. These proteins are the key regulators that sustain life. There are certain proteins that are shared among all living organisms from one end of the evolutionary spectrum (humans) to the other (bacteria), while others are more unique to a certain class of organism. Living cells are constantly dividing and, therefore, required to duplicate their genomic information (DNA) in order to sustain life. Hence, the proteins involved in catalyzing such essential cellular proccesses must conserve their functionality throughout evolution, while the less essential proteins are more unique to certain kingdoms, phyla, or classes of organisms. The DNA polymerase enzymes and the MHC (major histocompatibility complex) molecules are examples of such proteins. DNA polymerase is essential for DNA replication and, therefore, neccessary for all living organisms, while MHC molecules are responsible for antigen presentation to trigger an immune response and are exclusively found in higher eukaryotes, such as humans and rats.

Homologous proteins are generally referred to as polypeptides that share a similar amino acid composition. In most cases, the proteins with relatively high degrees of identity are also structurally and functionally homologous (Figure 5.6). Changes in the protein's amino acid sequence could result in a

---

* Color Figure 2 follows page 52.

change of its three-dimensional structure. It is this relationship between the protein's amino acid sequence and its three-dimensional structure that allows us to compare proteins that lack X-ray crystallography or NMR-resolved structures with their sequence homologs having known structures.

FROM DNA SEQUENCE TO FUNCTION

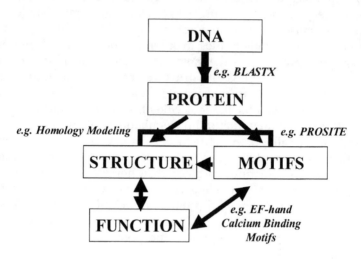

FIGURE 5.6
From DNA sequence to protein function

The sequence homologs with known structures allow us to calculate the structure of the homologous sequences that lack structures by comparative modeling and, thereby, to gain insight into the protein's function. In homology-based protein modeling, the experimentally determined structures are generally referred to as the "templates" and the sequence homolog (e.g., a novel string of nucleotides identified from ongoing genome projects) that lacks structural coordinates is called the "target" sequence. The homology-based protein modeling approach entails four sequential steps (Figure 5.7); the first involves the identification of known structures that are related in sequence to our target sequence. This step is typically achieved by using Internet tools such as BLAST to search for the potential templates. In the second step, these potential templates are aligned with our target sequence to identify the closest related template. In the third step, a model of the target sequence is calculated from the most suitable template found in step two. The final step involves the evaluation of our modeled target sequence using a variety of criteria (e.g., energetics, etc.).

Ideally, the amino acid composition of the unknown protein and its known structural homologs are quite similar and more than one known structural

Homology-based Modeling of Structures

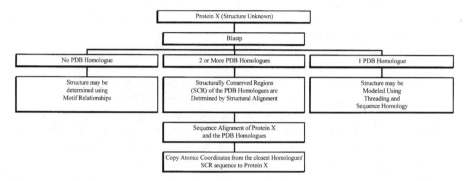

FIGURE 5.7
Homology-based modeling of structures

homolog from a different species is present in the PDB[1] protein database. NCBI's BLAST[2] enables us to find these known homologs. BLAST uses the primary sequence (amino acid sequence) of the target sequence as its input file and conducts a search for homologous protein sequences in a defined database (e.g., SWISS-PROT,[3] PDB, etc.). Upon completion, the closest sequence homologs are displayed. The related proteins are generally those with the highest BLAST scores. By defining the database as PDB, the search would be limited to the sequence homologs that have known structures. The structural coordinates of the highest scoring proteins are then retrieved and used as input files in the protein modeling program of interest (e.g., MSI's homology software). The outcome of such a structural modeling of a consensus structure is shown for serine proteases in Color Figure 3*. The core structure of four different proteases is highly conserved (yellow) which includes the catalytic center. The surface loop structures are colored red and blue indicating their degree of deviation from the four input structures. The process of obtaining this consensus structure is described in the three subsequent figures: sequence alignment in Color Figure 4,* loop identification in Color Figure 5,* and structure overlay in Color Figure 6* (not a consensus structure).

The atomic coordinates of the potential templates are then structurally aligned to display the structurally conserved regions (SCRs) of the protein group of interest. The use of SCRs will enable the construction of the evolutionary conserved structural features of the target sequence (structurally unknown protein sequence). Although there seems to be a relationship between the protein's amino acid sequence and its structure, the ambiguities involved in this relationship currently prevent us from constructing models of structurally unknown proteins based solely on their amino acid sequence.

---

* Color Figures 3, 4, 5, and 6 follow page 52.

The knowledge of evolutionarily conserved structural features of similar proteins from other species thus enables us with greater confidence to gain insight into the structure of the target sequence. Regions of the protein that are highly homologous in their amino acid sequence do not always correlate to a structurally conserved region (SCR). In fact, many of the loop regions of homologous proteins are highly similar in their amino acid sequence, but their structural features are quite variable (Color Figure 5). In other words, the relationship between the SCRs and their respective amino acid sequence alignment is not as obvious as would be expected. Therefore, the presence of the SCRs is essential in such a comparative approach. After the assigment of the SCRs, the most suitable template must be found. The template is typically the closest amino acid sequence homolog to the target sequence. This usually happens to be the closest species to our target sequence. If our target sequence is protein X from human, and the potential templates are protein X from rat and *Drosophila* (fly), then on the basis of phylogeny, the rat protein X is typically selected over *Drosophila* and is used as the template sequence for modeling the human target sequence (protein X). This is based on the assumption that human and rat are the closer sequence homologs.

The amino acid sequence of the target protein is then aligned with the chosen template sequence (Color Figure 4). The atomic coordinates of the amino acids in the SCR regions of the template sequence are copied over to their respective sequence homologs in the target protein. Most of the secondary structures (e.g., alpha helix, beta strand, etc.) are typically covered by the SCRs. The structural features that are typically outside of the labeled SCRs are the loops and the terminal ends of the molecule. After assigning coordinates to the SCR overlapping amino acid sequences, the most suitable structural coordinates for the target protein's loop regions and its terminal ends must then be found.

The following section summarizes a reliable, scientifically sound, and extremely user-friendly set of protein structure prediction-and-analysis software available from Molecular Simulations Incorporated (MSI). MSI (www.msi.com) offers proprietary computational tools that are widely used by academia and are also quite popular in industry.

Structure Prediction Software

Structure prediction tools are generally classified into two categories: public domain and proprietory software. Following is a summary of some of the many proprietary software tools exclusively offered through MSI:

What is proprietary software (e.g., MSI software)?

These are tools that must be purchased by those who maintain the rights to the software. Their popularity is based mainly on their leading-edge approach to predicting and analyzing structural characteristics of the macromolecules and their sequence homologs. These proprietary algorithms are very popular in the pharmaceutical field, and their use has revolutionized the study of drug design by adding an element of prediction to the old empirical approach. One of today's most popular proprietary programs is the InsightII environment offered by MSI (it was used to generate some of the figures shown in this book). MSI is a biotechnology and bioinformatics company based in La Jolla, California specializing in structure prediction and analysis software. MSI is predominantly involved in designing and maintaining its own proprietary algorithms and software that meet the needs of a variety of biotechnology and pharmaceutical companies. The company also collaborates with many academicians and constantly strives to sustain its technological leading edge. MSI provides a variety of user-friendly yet sophisticated software through the InsightII environment, as well as software specifically designed to perform delicate computational tasks. For instance, the tools in the Affinity software program can be used in rational drug design, while tools such as Homology are utilized to predict the three-dimensional structure of a protein through sequence homologs with known 3-D structures.

What type of hardware platform is required for the MSI software?
Currently, MSI's InsightII environment is only available for Silicon Graphics and IBM RISC system/6000 workstations.

What is the InsightII graphics environment?
Most of the MSI software modules require the InsightII 3-D graphics program. Its graphics environment provides a very user-friendly setting (as shown in Color Figure 7*) that facilitates the use of all other compatible software modules. A variety of interactive tutorials are also provided for many of their sophisticated computational tools to further facilitate the learning process.

Following is a list of some of the MSI modules currently utilized in academia and industry (adapted from the MSI homepage: www.msi.com):

- Biopolymer: this module helps to construct protein, nucleic acid, and carbohydrate models. These models can then be used for a variety of simulation jobs.
- Profiles – 3-D: this module enables the user to evaluate the compatibility of the amino acids in their environment to a database of 3-D profiles. The compatibility is a verification measure of the modeled structure and its sequence[1]. This method tries to answer the following question: is the query sequence compatible to a known three-dimensional profile?
- DelPhi: this module is typically used to calculate solvation energies and electrostatic potentials.

---
* Color Figure 7 follows page 52.

- Homology: this module is used to construct 3-D models of proteins using sequence homologs with known structures as templates (Color Figures 3–6).
- ELF: this module enables the user to calculate the minimum structures and various molecular interactions. It can also be used to calculate solvent-induced effects. These effects include both electrostatic and binding-site interactions.
- Discover: this module is typically used for energy minimization, dynamic simulations, and conformational sampling. It incorporates a variety of validated force fields. Discover enables the user to predict a variety of energetic and structural properties associated with the macromolecule or system of interest (Color Figure 2).
- CHARMm: this is a somewhat more advanced module that combines minimization and dynamics features with expert methods such as free energy perturbation (FEP), combined quantum and molecular mechanics (QM/MM), or correlation analysis.
- Affinity: this is an automated docking program. The module is typically used in rational drug design. It docks the appropriate LIGAND into its receptor.
- Ludi[4]: this module allows the investigator to fit the LIGAND or molecule of interest into the active site or binding site of the targeted protein. It accomplishes this by matching the complementary nonpolar–nonpolar and polar–polar substituents.

For a more detailed explanation of MSI's software and services, please visit their home page at www.msi.com. Many of the color figures in this volume were created using MSI software. Special thanks to our colleagues at MSI for providing us with these educational and visually stunning pictures.

References

1. Sussman, J.L., et al., Protein Data Bank (PDB): database of three-dimensional structural information of biological macromolecules. *Acta. Crystallogr. D. Biol. Crystallogr.*, 1998. 54(1 (Pt 6)): p. 1078-84.
2. Altschul, S.F., et al., Basic local alignment search tool. *J. Mol. Biol.*, 1990. 215(3): p. 403-10.
3. Bairoch, A. and R. Apweiler, The SWISS-PROT protein sequence data bank and its supplement TrEMBL in 1999. *Nucleic Acids Res.*, 1999. 27(1): p. 49-54.
4. Bohm, H.J., LUDI: rule-based automatic design of new substituents for enzyme inhibitor leads. *J. Comput. Aided Mol. Des.*, 1992. 6(6): p. 593-606.

Glossary

Amino acid.	Small molecules with various chemical properties forming the building blocks of → *proteins*.
Base.	Distinct chemical structures found in → *nucleic acids* and part of → *nucleotides*. The bases of nucleotides form the signature letters allowing sequence information to be stored in → *DNA* and → *RNA* strands.
Bioinformatics.	Computational analysis of biological information such as → *nucleic acid* and → *protein sequences* and *protein structures*.
cDNA.	Complementary DNA obtained from a messenger RNA template through a process called reverse transcription.
Cell.	Basic, self sustaining unit of living organisms. Composed of cell → *membrane* as outer boundary surrounding the → *cytoplasm* and internal → *organelles* that carry out specialized functions (see → *mitochondrion*).
Cell culture.	Artificially (in vitro) maintained cell population in growth medium containing specifically isolated cell types which grow indefinitely; used to express → *recombinant DNA* or → *proteins* for physiological studies simulating experiments that would have been done in living organisms.
Chromosome.	Structurally independent unit of a → *genome* (see also *karyotype*).
Cloning, clone.	1. Organism: reproducing a genetically identical offspring; 2. Gene: duplicating a → *nucleic acid sequence* without introducing → *mutations*.
Contig.	Contiguous → *DNA sequence* obtained from individually sequenced DNA fragments that contain overlapping sequences at their ends.
Cytoplasm.	Content of cell in which most metabolic processes occur.
DNA.	Deoxyribonucleic acid; is part of chromosomes and contains the genetic information in all organisms.
Enzyme.	Protein that catalyzes a chemical reaction.
Eukaryote.	Group of organisms that contain organelles within their → *cytoplasm*, specifically a nucleus containing all → *chromosomal material*.

Evolution. (Biological evolution) the perpetual change of the genetic composition of living organisms.

Exon. Gene sequence on chromosome that belongs to the coding sequence; exon sequences are interrupted by → *introns*.

Gene. Hereditary unit of life on → *DNA* in → *chromosomes*; individual genes code for → *proteins* or → *RNA molecules*; in → *eukaryotic cells*, most genes are structured in → *exon* and → *intron* structures with the former containing the coding sequence.

Gene pool. Collection of all genes or coding sequences within a population of an organism.

Genetic code. The "language" by which genetic information is stored on chromosomes. Consists of four "letters" or bases: A (adenine), G (guanine), C (cytosine), and T (thymine) where triplets form codons; each codon represents an amino acid.

Genome. Total genetic content of an organism, both structurally and functionally.

Genotype. Hereditary unit of the → *genome* (usually a → *gene* or group of genes).

Homology. Evolutionarily derived similarity between genes or proteins.

Intron. Chromosomal DNA sequences interrupting the coding sequence of genes (called → *exons*).

Karyotype. Chromosome set of an organism, species specific; chromosomes are arranged by size and characterized by their banding patterns.

Library. 1. DNA library: collection of DNA fragments of various origin; 2. Chemical library: collection of compounds generated by random, combinatorial synthesis strategies.

Membrane. Semi-permeable cell envelope made of phospholipids and membrane proteins; while phospholipids provide stability of membranes, proteins provide transport and signaling processes across this otherwise impermeable structure.

Metabolism. Chemical reactions in → *cells* for the degradation and biosynthesis of molecules. Chemical energy is extracted from nutrients and used to synthesize macromolecules, promote transport, signaling, and growth.

Mitochondrion. Organelle in → *eukaryotic organisms* responsible for oxygen-dependent energy → *metabolism*.

Mutation. Change in the → *base* or → *nucleotide* sequence in a gene or → *chromosomal* structure.

Nanometer, nanotechnology. One nanometer is one billionth of a meter; nanotechnology pertains to molecular devices with dimensions in the nanometer range.

Neuron.	Nerve cell in the brain responsible for electrical and chemical signal transmission.
Nucleic acid.	Macromolecule important for storage of genetic information; is composed of → *nucleotides* which determine the sequence of → *genes*; comes in two common forms, → *DNA* (deoxyribonucleic acid) and → *RNA* (ribonucleic acid).
Nucleotides.	Building blocks of → *DNA* and → *RNA*; composed of → *base*, ribose (sugar), and phosphate groups.
Organelle.	Small → *membrane*-encapsulated particle in the cytoplasm of → *eukaryotes.*
Ortholog.	A homologous sequence that is derived from a common ancestor found in individuals of different species.
Paralog.	A homologous sequence that is derived from gene duplication and found within the same organism.
PCR, polymerase chain reaction.	An enzyme-mediated DNA amplification mechanism which allows sequence-specific selection of the DNA to be amplified.
Polymorphism.	Sequence differences among individuals found on specific → *chromosome* locations within a population.
Protein.	Macromolecules that carry out most functional activities in cells (cell structure, regulation, digestion, biosynthesis); classes of proteins include enzymes, hormones, → *receptors*, and antibodies.
Receptor.	Protein that serves as binding site for signaling molecules such as growth factors, hormones, or neurotransmitters.
Recombinant DNA.	Genetically modified or structurally altered → *nucleic acid* sequence (usually a → *gene*) within a → *vector* or host → *DNA.*
RNA.	Ribonucleic acid; a form of → *nucleic acid* that comes in three distinct polymeric types: messenger RNA which mediates the translation of DNA into amino acid sequences; transfer RNA which couples amino acids with the corresponding → *codon* on the messenger RNA; rRNA which is part of ribosomes, the cytoplasmic particles catalyzing → *protein* biosynthesis.
Vector DNA.	Small piece of → *DNA* containing regulatory and coding sequences of interest; vector DNA functions to insert and amplify a gene into a target → *genome.*

Index

2

2-D gels 133
  analysis software 134
  caveat 134
  databases 133
2-D imagery, drawbacks 158
2-D page
  databases 136
  public domain analysis tools 122

3

3-D visualization 154

A

Acidic residues 18
Action potentials 149
Adenine 28
Alanine 18
  hydrophobicity 20
Algorithms Project 146
Alignment tools 42
Allele 86
  information 51
  variations 110
Alliance of Genetic Support Groups 43
Alpha amino acids 19
Alternative yeast nuclear code 103
Alzheimer's 150
Amino acid 17
  defined 169
  peptide bonds 19
  sequencing 94
Anatomical phenotypes 111
Anatomy, visualization 157
Arabidopsis thaliana 5
Archaea 31, 32

Arginine 18
  hydrophobicity 20
Ascidian mitochondrial code 103
Asparagine residues 19
  hydrophobicity 20
Aspartate 18
  hydrophobicity 20
Astigmatism 87
Avery, Oswald 28

B

Bacterial code 103
Bacterial genomes
  screening with ORF finder 103
Bacterial promoter 84
Bacteriology 15
BankIt 38, 43
Base sequences 137
Base, defined 169
Basic BLAST entry form 99
Basic residues 18
Bias, in data analysis 13
Bibliographic searches 39
Binary relations 137
Binding sites 80
Biochemistry 15
Bioinformatics
  beginning of 28
  defined 2, 169
Biological data management
  system 15
Biological function, learning from
  sequence 101
Biological macromolecules 17
  generating images 157
Biological motifs 80
Biological transporters, protein
  function 26

Voxel model 158

W

Watson, James 28
Web sites, limitations and reliability
    89
Whole Brain Atlas 145
Wilson's disease 153
Wiring diagrams 137

X

X-linked Charcot-Marie Tooth
    disease 153
Xonobiotics, expression profiles 136
X-ray crystallography 161

Y

Yeast
    electrophoretic information
        134
    genome project 136
    mitochondrial code 103
    protein databases 107
Yeast Proteome Database 136

Z

Zellweger syndrome 153